WIRELESS TRANSCEIVER DESIGN

WIRELESS TRANSCEIVER DESIGN

Mastering the Design of Modern Wireless Equipment and Systems

Ariel Luzzatto and Gadi Shirazi

John Wiley & Sons, Ltd

Other Wiley Editorial Offices

John Wiley & Sons Inc., 111 River Street, Hoboken, NJ 07030, USA

Jossey-Bass, 989 Market Street, San Francisco, CA 94103-1741, USA

Wiley-VCH Verlag GmbH, Boschstr. 12, D-69469 Weinheim, Germany

John Wiley & Sons Australia Ltd, 42 McDougall Street, Milton, Queensland 4064, Australia

John Wiley & Sons (Asia) Pte Ltd, 2 Clementi Loop #02-01, Jin Xing Distripark, Singapore 129809

John Wiley & Sons Canada Ltd, 6045 Freemont Blvd, Mississauga, ONT, L5R 4J3, Canada

Wiley also publishes its books in a variety of electronic formats. Some content that appears
in print may not be available in electronic books.

British Library Cataloguing in Publication Data

A catalogue record for this book is available from the British Library

ISBN: 978-0-470-06076-6 (HB)

Typeset in 11/13pt Times by Laserwords Private Limited, Chennai, India.
Printed and bound in Great Britain by Antony Rowe Ltd, Chippenham, Wiltshire.
This book is printed on acid-free paper responsibly manufactured from sustainable forestry
in which at least two trees are planted for each one used for paper production.

To my wife Efrat for her support and love
Ariel Luzzatto

To my wife Iris, and our children, Eran and Omer
for their support and love
Gadi Shirazi

Contents

Preface

Background

The insatiable requirement for high-speed real-time computer connectivity any-where, at any time, fuelled by the wide-spreading acceptance of the Internet Protocol, has accelerated the birth of a large number of wireless data networks.

Buzzwords, such as WiFi, Bluetooth and WiMax, have already become every-day language even for people unfamiliar with their technological meaning. They all, however, refer to the same basic functionality: the transfer of high-speed data through wireless networks.

As we proceed in the twenty-first century, the variety of wireless standards is far from converging, since each one has its own peculiar advantages. Trying to figure out their evolution is very difficult. The only certain fact is that all of them will seek to enable digital communications through broadband wireless equipment, and one of the main tasks being the capability of allowing a large number of different users to coexist and operate in a crowded and often unregulated electromagnetic environment.

The design of modern digital wireless modems and transceivers, capable of supporting high-speed data protocols in such wild scenarios, is very different from the traditional one. Many of the components in the wireless chain require an integration scale whose cost can be justified only for extremely large production quantities, thus, their design and production is way beyond the capability of most hi-tech industries. As a consequence, as happened with digital processors and memories, R&D engineers must now learn how to manage using off-the-shelf multi-purpose components manufactured by a few giant chipmakers.

In contrast, several of the most critical subsystems, such as voltage controlled oscillators (VCO), linear power amplifiers, fast-hopping synthesizers etc., are so diversified and application-dependent, that in many cases there exists no suitable components from standard lines of products.

The above technological trends are developing much faster than the available literature, so that industrial R&D engineers, as well as academic researchers and lecturers, often have difficulties in accessing the relevant technical knowledge. Therefore, while taking a comprehensive and self-contained approach, we place special emphasis on subjects of critical importance, whose theory and know-how is mainly dispersed in cryptic papers for professionals, and cannot often be found in an organized and easy-to-understand form.

Scope

This book is intended to be an academic-level, comprehensive, self-contained and friendly guide to theoretical and practical modern wireless modem and transceiver design for experienced radio and communication engineers, as well as providing a basis for advanced undergraduate and postgraduate engineering courses.

On the one hand, attaining a good academic level implies the use of advanced mathematical and theoretical tools, and on the other, for the sake of practical use, the reader should be capable of rapidly gaining in-depth understanding. Thus, we take special care to leave no 'blind spots' or mysteries, and, whenever a background topic is brought up for the first time, we give a short reminder.

While carefully watching out for technical rigor, we always keep an eye on the practical side, avoiding esoteric theoretical dissertations, and in parallel, we do not waste time with 'old stuff' that is of no use for modern designers. In the case of well-known and well-treated topics, we briefly explain the applicable results, and refer the reader to the literature for more insight, and place special emphasis on subjects of critical importance, which, we believe, are less commonly treated.

We show how to analyze and master critical parasitic phenomena that sometimes are thought to be out of the designer's control, however, if overlooked, may show up at a later stage leading to after-market disasters.

Clarifying and detailed computational examples are employed throughout the book, and analysis and design are carried out for each of the most useful modern wireless architectures.

The material is arranged so it may be consulted ad hoc, in other words, each chapter is self-contained, and whenever needed, reference to specific subjects in other chapters is given.

At the end of the day, the users of this book will master techniques and tools of long-term applicability, which will provide in-depth insight of topics otherwise hard to analyze and dominate. In particular the reader will:

- Fully understand the specifications that characterize the performance of modern wireless modems and transceivers as a whole system, learn how to measure them, and discover how they depend on (one or more) components and subsystems in the various architectures of widespread use.
- Understand the important figures related to off-the-shelf radio-frequency and baseband super-components, which often are not explicitly published, or are defined with foggy descriptions, and learn how to measure them to fully evaluate applicability and limitations.

- Understand and practice the detailed design of these circuits and subsystems, which usually cannot be purchased off-the-shelf, and are not as fully covered in the literature.
- Learn to master the design, analysis and measurement of important and hard-to-achieve parameters, such as phase noise and spurs of oscillators and synthesizers, peak-to-average and linearity of radio-frequency power amplifiers, radiated spurious emission.
- Learn how to analyze and dominate critical parasitic phenomena such as microphonics, remodulation, oscillator leakage, parasitic phase-locked-loop (PLL) poles, reference spurs, high-frequency power amplifier instability – how they affect the transceiver performance, how to measure them and minimize their impact.

About the Authors

Ariel Luzzatto has 30 years of experience, most of them designing commercial and industrial communication and RF products with Motorola Israel, and is a former lecturer of communication circuits and systems with Tel Aviv University. He holds a Ph.D. and a M.Sc. in applied mathematics, and a B.Sc. in electronic engineering from Tel Aviv University.

Gadi Shirazi has 25 years of experience, most of them designing advanced communication and RF products with Motorola Israel, and is a former lecturer of communication circuits and systems with Tel Aviv University. He holds a M.Sc. and a B.Sc. in electronic engineering from Tel Aviv university, and 24 patents in the field of communications.

Abbreviations

A/D	analog to digital
ACPR	adjacent channel power ratio
ADSL	asymmetric digital subscriber loop
AM	amplitude modulation
BER	bit error rate
BPF	band-pass filter
BPSK	binary phase shift keying
BT	Bluetooth
BW	bandwidth
CCR	co-channel rejection
CDF	cumulative distribution function
CDMA	code division multiple access
CMOS	complementary metal oxide semiconductor
DC	direct current
DCR	direct conversion receiver
DDFMT	dual port FM transmitter
DFMT	direct FM transmitter
DLT	direct launch transmitter
DSP	digital signal processor
ESR	equivalent serial resistance
ET	envelope tracking
EVM	error vector magnitude
FDMA	frequency division multiple access
FDTD	finite-differences time-domain
FH	frequency hopping
FM	frequency modulation
GMSK	Gaussian minimum shift keying
GSM	global system for mobile communication
HIFR	half-IF rejection
IF	intermediate frequency
IMD	intermodulation distortion
IMR	intermodulation rejection

IP	intercept point
IQ	inphase and quadrature
LNA	low noise amplifier
LO	local oscillator
LPF	low-pass filter
MASH	multi-stage noise shaping
MIMO	multi-input multi-output
NF	noise figure
NRZ	non-return to zero
OFDMA	orthogonal frequency division multiple access
PA	power amplifier
PAPR	peak-to-average power ratio
PDF	probability distribution function
PLL	phase lock loop
PM	phase modulation
QAM	quadrature amplitude modulation
QPSK	quadrature phase shift keying
RF	radio frequency
RSSI	received signal strength indicator
SAR	specific absorption rate
SHR	superheterodyne receiver
SINAD	signal + noise + distortion over noise + distortion
SNR	signal-to-noise ratio
SoC	system-on-chip
SSB	single sideband
TDD	time division duplex
TDMA	time division multiple access
TETRA	terrestrial trunked radio
TSCT	two-step conversion transmitter
UUT	unit under test
UWB	ultra-wideband
VCO	voltage-controlled oscillator
VLIF	very low IF
VSA	vector signal analyzer
WLAN	wireless local area network

1

Modern Transceiver Architectures

Modern transceiver architectures are very different from traditional ones. Many of the functions traditionally belonging to radio-frequency (RF) circuitry have been taken over by digital signal processors, and the boundaries between baseband (low-frequency) functionality and radio-frequency performance have become fuzzy.

Several different approaches have taken the lead, each with its peculiar advantages and drawbacks. One aim of this chapter is to gain a general insight into the important features of the various transceiver architectures, show how they can be implemented with off-the-shelf RF and baseband components and understand their applicability and limitations.

Another goal is to gain an understanding of what the various transceiver specifications mean, how they influence the total RF system capability, how they are affected by the architecture of choice and by the various transceiver subsystems and their mutual interaction.

Once we are able to accurately define the subsystems, the transceiver design reduces to pick-up the proper off-the-shelf components and carrying out a detailed design where no ready-made parts are available.

1.1 Overview

The small physical size and low cost, required by competitive commercial equipment, together with mandatory multisystem operability and very low current consumption, dictate large integration scales for both baseband and RF subsystems. Such a high integration level on a chip cannot be achieved with conventional transceiver configurations.

Traditional discrete-component constructions must be abandoned in favor of architectures where ultra-fast and specialized digital signal processors (DSP) are responsible for most transceiver functionality and performance, while allowing improved spectral efficiency and utilization.

However, in most cases, the cost of designing and manufacturing dedicated chips can be justified only for extremely large production quantities.

Wireless Transceiver Design   Ariel Luzzatto and Gadi Shirazi
© 2007 John Wiley & Sons, Ltd

It follows that, just as digital designers must rely on off-the-shelf processors, memories and peripherals, so too must RF designers adopt system-on-chip (SoC) multi-purpose components designed and manufactured by a few giant chipmakers, which are often not optimized or even meant for carrying out their actual task.

On one hand, the chronic lack of insight regarding SoC performance and the risk of discovering hidden pitfalls at later stages, require the designers to carry out a thorough evaluation of chip parameters, which often are not explicitly published or are only loosely defined.

On the other hand, several of the most critical RF subsystems to be operated in conjunction with an SoC are so diversified and application-dependent, that often no standard line of products fits, and they must be designed ad-hoc.

1.2 Receiver Architectures

We can arbitrarily divide a receiver into three main functional blocks:

- Front end: all the circuits that carry out operations at final RF frequency, such as RF front filters, low-noise amplifiers (LNA), high frequency mixers etc.
- Intermediate frequency (IF) chain: all the circuits operating at non-zero IF frequency (if any).
- Backend: all the circuits operating at a frequency below first IF (if any) or other than final RF frequency, such as baseband processing, detector etc.

The critical parameter that we focus on, in determining the performance of a receiver, is the signal-to-noise ratio (SNR) at the input of the detector (usually at the input of the digital baseband sampler in the backend). By SNR we mean, in a broad sense, the ratio between the desired signal power and the cumulative effect of all the unwanted phenomena.

We always translate every disturbance (noise, on-channel and off-channel interferers etc.) and every receiver limitation (nonlinearity, noise figure (NF), synthesizer spectral purity etc.) into an equivalent SNR at the detector input, and the design of all the blocks preceding the detector will aim to obtain a certain SNR value.

Then, the detector transfer function under the given SNR will determine the received signal quality for a given modulation method (i.e., bit error rate (BER) for digital modulations, or signal + noise + distortion over noise + distortion (SINAD) for analog ones).

Other important design considerations, together with small physical size and low cost, will be functional requirements such as low current consumption, fast wake-up time, multi-band, multi-system flexibility etc.

The two most important receiver (Rx) architectures on which we focus in this book are the modern superheterodyne receiver (SHR), useful for most high-tier broadband and narrowband applications (Rx specs of the order of 75 dB and up), and the direct conversion receiver (DCR), for mid-tier broadband applications (Rx specs ranging from 50 dB to 75 dB).

Both usually utilize a split backend architecture including inphase and quadrature (IQ) baseband channels. With certain modulation types, SHR can also work in single-ended configurations.

A third, and less common approach, the very low IF (VLIF) receiver, serves narrowband low-tier and low-cost equipment (Rx specs of the order of 50 dB or less), which often uses sophisticated versions of IQ channels.

For each architecture, we assign a table with typical standalone specifications for every subsystem in the receiver. The values in the table will be used in Chapter 2 to compute all the different Rx specifications.

We do not include here well-known topics on receiver functionality, however, we believe that the examples given in Chapter 2 will suffice to refresh the theory of operation, will clarify and illustrate the mechanisms governing the various phenomena and show the relationships between the individual building blocks and the global RF performance.

1.2.1 Superheterodyne Receiver (SHR)

The SHR with IQ backend is the right choice for delivering high-performance narrowband and broadband receiver specs. Its configuration is illustrated in Figure 1.1.

Figure 1.1 shows the operational frequency of each block, where f_R denotes the desired receiver frequency, f_{IF} the IF frequency, and bb denotes baseband frequencies.

The baseband output is digitally sampled and handed over to a baseband DSP in order to perform detection and further processing on the received signal. As far as the RF design is concerned, however, we do not need to know a priori what happens after the digital sampling. All we have to care about is the SNR at the output of the IQ low-pass filters (LPF).

Then, for a given SNR level, the output performance is a function of the modulation type, coding scheme etc., which may change dynamically if the radio has multimode functions.

The SHR as well as the IQ principles are well known, thus we describe them briefly here, and we refer the interested reader to the detailed analysis of Chapter 2 and to the Bibliography for more detail.

For computation purposes, we assign, in Table 1.1, typical specifications for each one of the building blocks of Figure 1.1 for a high-performance public-safety receiver in the 850–870 MHz band, with 25 kHz channel spacing.

1.2.1.1 SHR Highlights

The received signal entering the antenna is filtered by the first band-pass filter (BPF) and second BPF (sometimes called 'preselectors'), whose function is essentially protection against various spurious phenomena, mainly far-out interferers

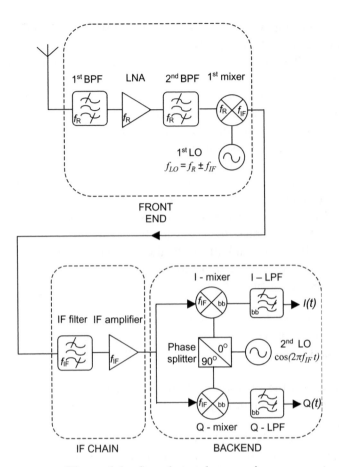

Figure 1.1   Superheterodyne receiver

(many channels away from the desired signal). The low-noise amplifier (LNA) contributes to sensitivity, but is not mandatory, and may be omitted when an active-type first mixer is employed.

The first mixer converts the incoming signal to IF frequency with the help of the local oscillator (LO) signal, often referred to as the 'injection'. In essence, the mixer performs the mathematical multiplication of the incoming signals with the injection signal, which ultimately results in subtracting (or summing) the instantaneous injection phase from the instantaneous phase of any carrier reaching the mixer input.

As we will see in Chapter 2, if the injection is corrupted by parasitic phase modulation, as a result of the phase subtraction, every incoming signal acquires the very same phase modulation of the LO. Therefore, if a strong interferer reaches the mixer input, it may spread out into the receive band, generating a serious disturbance.

Table 1.1 Typical subsystem values for a high-tier superheterodyne receiver

	First and second BPFs	LNA	First mixer	First LO	IF filter	Backend
Center frequency (MHz)	860	860	860	$f_R \pm f_{IF}$	73.35	0
Pass-band BW	19 MHz				18 kHz	2 × 9 kHz
Pass-band loss (dB)	1.5				3	1
Atten. @ $f_R \pm 2f_{IF}$ (dB)	35					
Atten. @ $f_R \pm 1/2f_{IF}$ (dB)	10					
Atten. @ stop-band (dB)	45				35	60
Atten. @ >9 kHz off. (dB)					40	60
Atten. @ $f_{IF} \pm 500$ kHz (dB)					40	
Atten. @ $f_{IF} \pm 1000$ kHz (dB)					40	
NF (dB)		2	7			11
Gain (dB)		12	−7			70
IP2 (dBm)			28			
IP3 (dBm)		0	10			−20
SBN @ Adj. Ch.(dBc/Hz)				−125		−118
SBN floor (dBc/Hz)				−155		−145
IQ amplitude imbalance (dB)						0.5
IQ phase imbalance (deg)						3

Note: $f_{IF} = 73.35$ MHz; f_R = received channel frequency: 850.5 MHz $\leq f_R \leq 869.5$ MHz.

The IF filter has the main purpose of providing protection against close-in interferers (one to few channels away). The signal from the IF chain is split into two channels, designated as I (inphase) and Q (quadrature), which are down-converted to zero center frequency for efficient further processing, and for providing additional close-in protection. Each one of the down-converted I and Q channels (also called 'quadrature channels') is a baseband signal with bandwidth of not more than half the RF signal bandwidth. The split is necessary because it is not possible to detect a general IF signal (simultaneously modulated in both amplitude and phase) using a single-ended mode approach. The split IQ quadrature architecture, however, is not always mandatory, and a single-ended backend can do well for FM-only or AM-only signals, especially when narrowband voice operation is required. However, the flexibility and the control achievable with the IQ split, makes it desirable for most modern radio designs.

1.2.1.2 Quadrature (IQ) Modulation Highlights

Because of the central importance of this subject, we review its basic theory in detail. The most general sinusoidal RF signal $S(t)$ transmitted with angular carrier frequency ω, including arbitrary low-frequency (low with respect to ω) and real-valued amplitude and phase modulations $A(t)$ and $-\pi/2 \le \varphi(t) < \pi/2$, can be represented in the form:

$$S(t) = A(t)\cos(\omega t + \varphi(t)) = Re[A(t)e^{j(\omega t + \varphi(t))}] \tag{1.1}$$

It is easy to see that Equation (1.1) can be written in the equivalent form

$$S(t) = Re[S_c(t)] = s_r(t)\cos(\omega t) - s_i(t)\sin(\omega t)$$

$$S_c(t) = [A(t)e^{j\varphi(t)}]e^{j\omega t} = [s_r(t) + js_i(t)]e^{j\omega t}$$

$$s_r(t) = A(t)\cos(\varphi(t)), \quad s_i(t) = A(t)\sin(\varphi(t)) \tag{1.2}$$

where $s_r(t) + js_i(t) = A(t)e^{j\varphi(t)}$ is often referred to as the 'complex amplitude' or the 'baseband signal'.

If $s_r(t)$ and $s_i(t)$ are known, then $A(t)$ and $\phi(t)$ are readily recovered from Equation (1.2) as

$$A(t) = \sqrt{s_r^2(t) + s_i^2(t)} \cdot sign[s_r(t)] \tag{1.3}$$

$$\varphi(t) = tan^{-1}[s_i(t)/s_r(t)] \in [-\tfrac{\pi}{2}, \tfrac{\pi}{2}) \tag{1.4}$$

An equivalent representation for which $A(t) \ge 0$ may be obtained by restricting $-\pi/2 \le tan^{-1}(x) \le \pi/2$, in which case

$$A(t) = \sqrt{s_r^2(t) + s_i^2(t)} \tag{1.5}$$

$$\varphi(t) = tan^{-1}[s_i(t)/s_r(t)] + \tfrac{\pi}{2}(1 - sign[s_r(t)]) \tag{1.6}$$

Down-mixing the received RF signal $S(t)$ with two split-injection signals of identical amplitude at carrier frequency ω, one of them (I path) exactly in phase with the RF carrier and the other (Q path) exactly in quadrature (with $+90$ degrees offset), recovers $s_r(t)$ at the output of the I path, and $s_i(t)$ at the output of the Q path.

Indeed, referring to Equation (1.2) and Figure 1.1, setting the LO equal to $cos(\omega t)$, with $|_{LP}$ denoting low-pass filtering, and with x^* denoting the complex conjugate of x, the received complex signal $I_R(t) + jQ_R(t)$ is

$$I_R(t) + jQ_R(t) = \{S(t)[cos(\omega t) - j\sin(\omega t)]\}|_{LP}$$

$$= \{(Re[S_c(t)])e^{-j\omega t}\}|_{LP} = \{\tfrac{1}{2}[S_c(t) + S_c^*(t)]e^{-j\omega t}\}|_{LP}$$

$$= \{\tfrac{1}{2}[s_r(t) + js_i(t)] + \tfrac{1}{2}[s_r(t) - js_i(t)]e^{-j2\omega t}\}|_{LP}$$

$$= \tfrac{1}{2}s_r(t) + j\tfrac{1}{2}s_i(t) = \tfrac{1}{2}S_c(t)e^{-j\omega t} \tag{1.7}$$

Equation (1.7) states that the recovered I and Q paths are perfectly isolated, i.e., if $s_i(t) = 0$, no signal will appear at the Q output due to $s_r(t)$ and vice versa. However, if the LO is not perfectly phase-locked to the carrier of the received signal, mutual coupling will occur. In other words, a distortion signal dependent on $s_r(t)$ will appear at the output of the Q channel together with $s_i(t)$ and vice versa.

A similar effect will result also when the I and Q injection signals are not mutually shifted by exactly 90 degrees, or the amplitude gain of the two quadrature paths is not identical.

Consider the simple, but very frequent case, where the LO of the receiver is out of phase, by a small angle ϕ, with respect to the carrier of the signal $S(t)$ in Equation (1.1).

Since ϕ is small, with a computation similar to Equation (1.7), the corrupted output $\tilde{I}_R(t) + j\tilde{Q}_R(t)$ can be expanded in a first-order Taylor series

$$\tilde{I}_R(t) + j\tilde{Q}_R(t) = \tfrac{1}{2}S_c(t)e^{-j(\omega t + \phi)} \approx [\tfrac{1}{2}s_r(t) + j\tfrac{1}{2}s_i(t)](1 - j\phi)$$

$$= [I_R(t) + \tfrac{\phi}{2}s_i(t)] + j[Q_R(t) - \tfrac{\phi}{2}s_r(t)] \qquad (1.8)$$

which shows that a parasitic cross-channel coupling of order $O(\phi)$ occurs due to the effective 'rotation' of the real and imaginary axes by ϕ. This effect, however, is easily compensated by de-rotating the axes a posteriori or by phase shifting the LO, based on a known training signal.

In the general case, where the I and Q injection signals have arbitrary small phase deviations θ and ϕ, and the I and Q channels have arbitrary small amplitude gain deviations ε and δ from nominal, we may write the corrupted received output $\tilde{I}_R(t) + j\tilde{Q}_R(t)$ in the form

$$\tilde{I}_R(t) + j\tilde{Q}_R(t) = \{\tfrac{1}{2}[S_c(t) + S_c^*(t)]$$

$$\times [(1 + \varepsilon)\cos(\omega t + \theta) - j(1 + \delta)\sin(\omega t + \phi)]\}|_{LP} \quad (1.9)$$

with a first-order Taylor expansion for small angles, and after some manipulation, the first-order approximation of Equation (1.9) yields

$$\tilde{I}_R(t) + j\tilde{Q}_R(t) \approx I_R(t) + jQ_R(t) + e(t)$$

$$e(t) = \tfrac{1}{2}[\varepsilon s_r(t) + \theta s_i(t)] + j\tfrac{1}{2}[\delta s_i(t) - \phi s_r(t)] \qquad (1.10)$$

Without going into fancy details, it is evident from Equation (1.10) that the deviation from the ideal and balanced values introduces an error signal of the order $O(\lambda)$, where $\lambda = max\{|\varepsilon|, |\delta|, |\theta|, |\phi|\}$, thus generating a 'noise' power of order $O(\lambda^2)$ relative to the desired signal.

It is customary to write the quadrature components in the form

$$A(t)e^{j\varphi(t)} = I(t) + jQ(t)$$

$$I(t) = A(t)\, cos(\varphi(t))$$
$$Q(t) = A(t)\, sin(\varphi(t)) \tag{1.11}$$

and to represent them as a time-dependent vector, called a 'phasor', describing some pattern in the complex plane as shown in Figure 1.2.

In the following we make use, without proof, of some of the concepts concerning digital modulation. A comprehensive treatment of the subject can be found in Sklar (2001).

In digital communication, specific phasors pointing at particular predefined locations in the complex plane are referred to as 'symbols'. Each symbol corresponds to a predefined stream of information, usually a combination of one or more bits. The collection of all the points in the complex plane corresponding to all the possible symbols belonging to the same modulation scheme is referred to as the 'constellation' of the scheme. As the constellation is allowed to be more dense, then each symbol carries more information. Figure 1.3 shows constellation examples.

In the modulation scheme known as 'binary phase shift keying' (BPSK) each symbol represents one bit of information; in 'quadrature phase shift keying' (QPSK) each symbol represents two bits, and in '16-point quadrature amplitude modulation' (16-QAM) each symbol is capable of representing four bits. Note that in both BPSK and QPSK all the symbols have the same amplitude and different

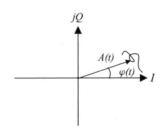

Figure 1.2   Representation of the quadrature components

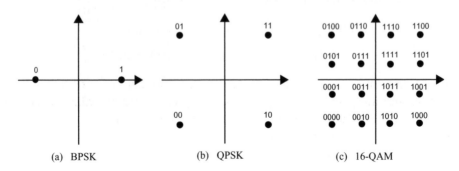

(a) BPSK   (b) QPSK   (c) 16-QAM

Figure 1.3   (a) 1 bit/symbol; (b) 2 bits/symbol; (c) 4 bits/symbol

phase, while in QAM each two symbols differ in amplitude and/or phase. Many different types of constellations with different topologies are possible.

Due to the influence of various parameters of the transmitter and the receiver (among them bandwidth and linearity) as well as the characteristics of the RF channel, when different symbols, each one of the same time duration, are sequentially transmitted, the phasors (the 'pointing' vectors) representing the complex amplitude, wander through the constellation describing a time-dependent pattern, which depends on the observation point along the transmit–receive (Tx–Rx) chain. In an ideal environment, regardless of the observation point, at certain equally spaced sampling instants, the phasors coincide exactly with one of the transmitted symbols following their original transmission sequence, thus by sampling the pattern anywhere in the Tx–Rx chain, the transmitted data may be exactly recovered. In practice, due to distortion and noise, the pattern described by the phasors does not cross exactly the locations of the symbols, but only passes nearby. In the detection process, the symbol which is closest to the phasor at the sampling instant, is considered the one that most likely has been transmitted. A typical record of the pattern generated by QPSK and 16-QAM symbols in a non-ideal band-limited environment looks as shown in Figure 1.4. The patterns are obtained by plotting $Q(t)$ vs. $I(t)$. Since the BPSK signal is one-dimensional it cannot be shown on a diagram of this type. A time segment of the corresponding $I(t)$ and $Q(t)$ signals for the 16-QAM case is shown in Figure 1.5.

Another graphic representation, known as the 'eye-pattern' provides an insight into baseband signal quality. The eye-pattern is well fitted for representing the BSPK case too, and is obtained by superimposing short segments of one of the time-dependent quadrature signals in synchrony with their common generating clock. The more 'open' the eye, the better the quality of the baseband signal,

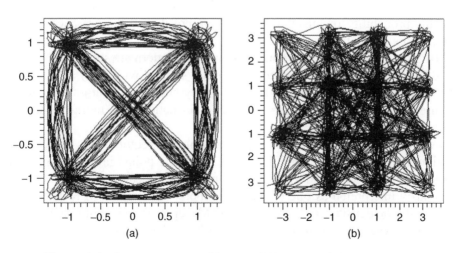

Figure 1.4 Patterns generated by: (a) QPSK; (b) 16-QAM symbols

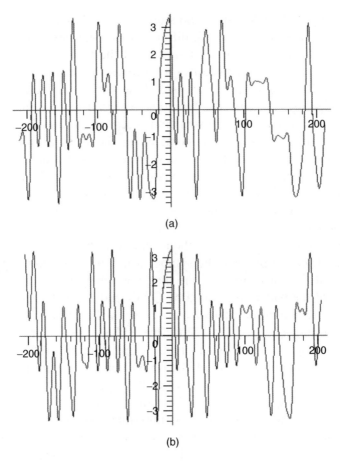

(a)

(b)

Figure 1.5   16-QAM quadrature channels vs. time: (a) $I(t)$; (b) $Q(t)$

since the probability of misinterpreting a symbol due to the presence of one of its neighbors is smaller.

A BPSK time signal and the corresponding eye-pattern are shown in Figure 1.6. The BPSK symbol may be viewed as a QPSK symbol with $Q(t) = 0$.

If the allowable bandwidth is fixed and the symbols switch at a fixed rate, then a 'denser' constellation may carry a higher information rate (more bits/second in a fixed bandwidth).

However, denser constellations put more stringent requirements on IQ channel isolation, SNR and group-delay distortion, since the points in the complex plane are closer to each other and a smaller disturbance may cause some symbols to be erroneously interpreted as one of their neighbors, resulting in BER degradation.

Constellation densities as high as 256-QAM, representing eight bits per symbol, can be used where excellent isolation, linearity and SNR are available.

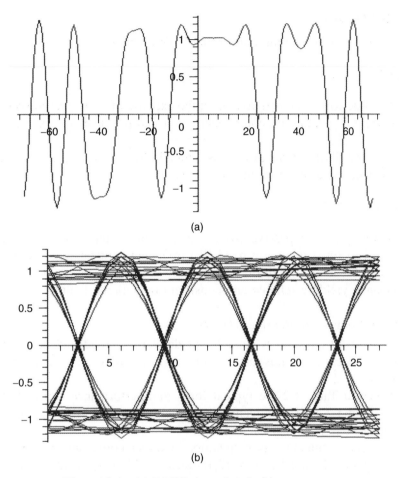

Figure 1.6   (a) BPSK time signal; (b) eye-pattern

The IQ approach has many advantages. Some of them are:

- A signal of bandwidth B is split into two signals of bandwidth at most $B/2$ each, thus each one of the I and Q channels may run at half the processing speed (of course, the overall bandwidth as seen at backend input before quadrature splitting is B).
- The IQ architecture is extremely suitable for generating and detecting high-speed digital modulation schemes.
- It is possible to formally work with complex signals, which makes it easy to treat one-sided Fourier spectra. This is of great help in understanding and simplifying signal handling. Also, the treatment of arbitrary signals is done in a unified and efficient way.

1.2.1.3 SHR Outlook

The following is worth noting concerning SHR architecture:

THE BAD NEWS

- High-level performance is delivered at the expense of an increased complexity, cost, current consumption, component count and physical size as compared to DCR and VLIF.
- Moreover the SHR architecture is less suitable for integration, due to the many chip connections to external lumped elements.

THE GOOD NEWS

- Due to the use of a passive front end, IF filters, and since the quadrature down-conversion is done at a fixed (IF) frequency, outstanding overall performance can be achieved with relatively mild constraints on the active backend components, yielding reliable and consistent operational figures.

1.2.2 Direct Conversion Receiver (DCR)

The DCR with IQ backend is suitable for delivering broadband data with mid-performance specs. It is illustrated in Figure 1.7, and typical specifications are given in Table 1.2.

The values in Table 1.2 are typical for a mid-performance WiFi (802.11b/g) RF modem in the 2.4 GHz unlicensed band, with about 20 MHz channel spacing and about 16 MHz occupied bandwidth.

The split IQ architecture is mandatory in direct conversion receivers, since mixing a general RF signal to zero-IF frequency with a single-ended architecture would produce non-recoverable aliasing (Schwartz, 1990).

A short description of DCR theory follows. For highlights on IQ theory see section 1.2.1.2 above.

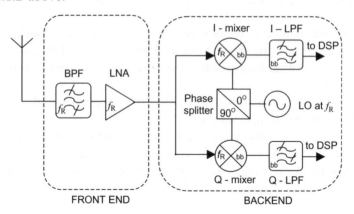

Figure 1.7   Direct conversion receiver

Table 1.2 Typical subsystem values for a direct conversion receiver

	BPF	LNA	Backend
Center frequency (GHz)	2.45		0
Pass-band BW (MHz)	83		8
Pass-band loss (dB)	1.5		
Attenuation @ stop-band (dB)	45		70
Attenuation @ > 8 MHz offset (dB)			70
NF (dB)		2	11
Gain (dB)		12	70
IP2 (dBm)			50
IP3 (dBm)		0	12
SBN @ Adj. Ch.(dBc/Hz)			−135
SBN Floor (dBc/Hz)			−135
IQ amplitude imbalance (dB)			0.5
IQ phase imbalance (deg)			3
Flicker noise corner @ BJT (kHz)		3	
Flicker noise corner @ CMOS (kHz)		500	

Note: $f_{IF} = 0$ MHz; $f_R =$ received channel frequency: 2.4 GHz $\leq f_R \leq$ 2.483 GHz.

1.2.2.1 DCR Highlights

The received signal entering the antenna is filtered by the front-end BPF whose function is mainly protection against various far-out spurious phenomena. The LNA is usually mandatory in a DCR, since the integrated backend mixers cannot easily deliver the noise figure (NF) required for sensitivity. The incoming signal is directly converted to quadrature baseband channels as described in section 1.2.1.2., with the help of the backend LO signal, which is right at the receiving frequency f_R. In absence of an IF filter, all the protection from close-in interferers must come from the I and Q low-pass filters.

Immunity from intermodulation as well as protection against various parasitic effects is inherent to the linearity, balance and quadrature accuracy of the I and Q channels. Moreover, parasitic phenomena arise from the fact that the LO frequency is identical to the received frequency. Thus the constraints on backend performance are often much more stringent than in the SHR case, as it can be appreciated comparing, for instance, the IP2 and IP3 (Chapter 6) specs in Table 1.1 to the ones in Table 1.2.

1.2.2.2 DCR Outlook

When using a DCR architecture we must be aware of:

THE BAD NEWS

- Backend linearity, spectral purity and IQ balance constraints are tougher than in the SHR case.

- The baseband frequencies near zero are prone to various interfering phenomena, such as '$1/f$' noise (Chapter 5), DC offset, self-mixing and LO leakage-generated Doppler interferers (Chapter 6). Thus, DCR baseband filters are usually built to 'notch-out' low frequencies. This 'DC hole' is not critical in broadband operation, since it covers only a small part of the overall bandwidth, but is often large enough to impair narrowband voice applications.

THE GOOD NEWS

- The DCR architecture greatly simplifies the receiver lineup, as can be appreciated comparing Figure 1.7 to Figure 1.1. A single LO is required, the front end can be simplified, and the IF chain is removed, leading to a major reduction in size and cost.
- Several hardware-driven tasks become software-driven, which makes it easy to dynamically adjust the channel bandwidth, and allows for flexible multimode operation.

1.2.3 Very Low IF Receiver (VLIF)

The VLIF receiver idea combines the advantages of both SHR and DCR approaches by using an IF frequency, usually ranging from half to several times the channel spacing Δf.

VLIF receivers have been implemented using many different architectures. We discuss here what we believe is the most flexible and easy implementation, and the one most likely to be found in off-the-shelf components.

The interested reader may find more insight in Crols and Steyaert (1995, 1998). For computation purposes we assume the typical specifications of a low-cost, hardware-based VLIF receiver to be similar to the values given for DCR in Table 1.2 except for the figures given in Table 1.3. Figure 1.8 illustrates a common VLIF architecture.

1.2.3.1 VLIF Highlights

If the low IF frequency is chosen to be at least equal to half the channel spacing, then two of the most critical receiver interferers, namely, adjacent channel (section 2.3) and image (section 2.6), are both filtered out by one complex BPF,

Table 1.3 Replacement values in Table 1.2 values for a very low IF receiver

	Backend
Attenuation @ stop-band (dB)	35
Attenuation @ > 8 MHz offset (dB)	35

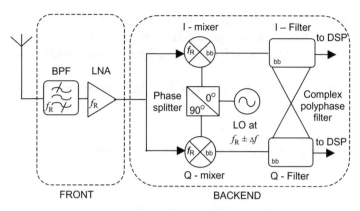

Figure 1.8   Very low IF receiver

usually referred to as a 'complex polyphase filter'. In the following we assume a low IF frequency equal to the channel spacing.

The complex polyphase filter performs a band-pass action, filtering the desired channel located at an IF frequency between $\Delta f/2$ and $3\Delta f/2$ with center frequency Δf, and cleaning out interferers. This filtering action cannot be accomplished by using low-pass filters on the I and Q channels, as previously done with the DCR approach. The reason is that upon mixing, interferers such as adjacent channel and image are mirrored in the frequency range between $-\Delta f/2$ and $\Delta f/2$, which is lower than the IF channel, and thus a low-pass filter will not do. Fortunately, it turns out that the required band-pass filtering can be performed by a simultaneous and mutual interaction between the I and the Q channels.

It is hard to see the results of such an interaction by looking at the IQ branches separately, but it is an easy task when looking at the signal in the complex domain, by using the 'complex amplitude' representation of section 1.2.1.2, as described next.

1.2.3.2 Complex Polyphase Filter Highlights

With $\omega_R = 2\pi f_R$ and $\Delta\omega = 2\pi\,\Delta f$, the received complex RF signal $S_c(t)$ is (see Equation (1.2))

$$S_c(t) = [s_r(t) + js_i(t)]e^{j\omega_R t} \qquad (1.12)$$

After down-mixing, say, with a lower-side LO injection $f_{LO} = f_R - \Delta f$, the resulting low-frequency complex signal $x(t)$, jointly defined by the real-valued signals $x_I(t)$ and $x_Q(t)$ at the input of the I and Q filters, is given by:

$$
\begin{aligned}
x(t) &= S_c(t)e^{-j(\omega_R - \Delta\omega)t} \\
&= [s_r(t) + js_i(t)]e^{j\Delta\omega t} \\
&= x_I(t) + jx_Q(t)
\end{aligned}
\qquad (1.13)
$$

where $s_r(t) + js_i(t)$ is band-limited to $\pm\omega_0$, $2\omega_0 \leq \Delta f$.

We note that the complex signal $x(t)$ in Equation (1.13), centered at Δf, is single-sided, i.e. it has no mirror at negative frequencies. Therefore, filtering $x(t)$ requires a complex, single-sided band-pass filter of bandwidth $2\omega_0$ centered at $\Delta\omega = 2\pi\Delta f$.

Denote by $H_{LP}(\omega)$, with $H_{LP}(-\omega) = H_{LP}{}^*(\omega)$, a conventional low-pass filter of bandwidth $\pm\omega_0$. Then the filter

$$H(\omega) = H_{LP}(\omega - \Delta\omega) \tag{1.14}$$

is a single-sided band-pass filter of bandwidth $2\omega_0$ centered at $\Delta\omega$.

Denote by

$$X(\omega) = X_I(\omega) + jX_Q(\omega) \tag{1.15}$$

the Fourier transform of the complex signal $x(t)$.

Filtering $X(\omega)$ with $H(\omega)$ yields a signal $Y(\omega)$ in the form

$$Y(\omega) = H(\omega)X(\omega) \tag{1.16}$$

where $Y(\omega)$ is jointly defined by the outputs of the I and Q channels, namely

$$Y(\omega) = Y_I(\omega) + jY_Q(\omega) \tag{1.17}$$

Note that the (complex) functions $X_I(\omega)$, $X_Q(\omega)$ and $Y_I(\omega)$, $Y_Q(\omega)$ must be Fourier transforms of real signals.

The question now is: how can one implement the filtering action of $H(\omega)$ by manipulating the real signals present at the edges of the I and Q channels? It is enough to show how to implement a single complex-pole transfer function, and then we may cascade similar stages to obtain more sophisticated filters.

The transfer function for a single-pole low-pass filter with 3 dB corners at $\pm\omega_0$ is:

$$H_{LP} = \cfrac{1}{1 + j\cfrac{\omega}{\omega_0}} \tag{1.18}$$

Then to obtain the single-sided band-pass, we use Equation (1.14) and get

$$H(\omega) = H_{LP}(\omega - \Delta\omega) = \cfrac{1}{1 + j\cfrac{(\omega - \Delta\omega)}{\omega_0}} = \cfrac{1}{1 + j\cfrac{\omega}{\omega_0} - j\cfrac{\Delta\omega}{\omega_0}} \tag{1.19}$$

Note that $H(\omega)$ has a complex pole. Substituting (1.19) into (1.16) yields

$$Y(\omega) = \left[X(\omega) + \left(j\frac{\Delta\omega}{\omega_0} - 1\right)Y(\omega)\right]\frac{\omega_0}{j\omega} \tag{1.20}$$

Assuming that no 'DC' signals are allowed in our system (which is the VLIF case), let us take the inverse Fourier transform of (1.20). With correspondence to the capital letters, the resulting time-domain complex equation is

$$y(t) = \omega_o \int_{-\infty}^{t} \left[x(\xi) + \left(j\frac{\Delta\omega}{\omega_0} - 1\right)y(\xi)\right]d\xi \tag{1.21}$$

To obtain (1.21) from (1.20), we used the fact that multiplication by $1/j\omega$ in the frequency domain corresponds to integration in the time domain.

Substituting into (1.21) the corresponding time-domain equations obtained by taking the inverse Fourier transform of equations (1.15) and (1.17), we may now split the real and imaginary parts (the split cannot be done in the Fourier domain because, in general, $X_I(\omega)$, $X_Q(\omega)$, $Y_I(\omega)$, and $Y_Q(\omega)$ are complex valued) getting

$$y_I(t) = \omega_0 \int_{-\infty}^{t} \left[x_I(\xi) - y_I(\xi) - \frac{\Delta\omega}{\omega_0} y_Q(\xi) \right] d\xi$$

$$y_Q(t) = \omega_0 \int_{-\infty}^{t} \left[x_Q(\xi) - y_Q(\xi) + \frac{\Delta\omega}{\omega_0} y_I(\xi) \right] d\xi \qquad (1.22)$$

or equivalently, in the Fourier domain

$$Y_I(\omega) = \left[X_I(\omega) - Y_I(\omega) - \frac{\Delta\omega}{\omega_0} Y_Q(\omega) \right] \frac{\omega_0}{j\omega}$$

$$Y_Q(\omega) = \left[X_Q(\omega) - Y_Q(\omega) + \frac{\Delta\omega}{\omega_0} Y_I(\omega) \right] \frac{\omega_0}{j\omega} \qquad (1.23)$$

which is readily manipulated into the form

$$Y_I(\omega) = \left[X_I(\omega) - \frac{\Delta\omega}{\omega_0} Y_Q(\omega) \right] H_{LP}(\omega)$$

$$Y_Q(\omega) = \left[X_Q(\omega) + \frac{\Delta\omega}{\omega_0} Y_I(\omega) \right] H_{LP}(\omega) \qquad (1.24)$$

The implementation of Equation (1.24) in the Fourier domain is shown in Figure 1.9. If the quadrature branches have a deviation of the order $O(\lambda)$ from the ideal values, with a reasoning similar to the one employed in section 1.2.1.2 one can roughly estimate the worst-case filter stop-band attenuation to be of the order of $20\log_{10}(\lambda)$ dB.

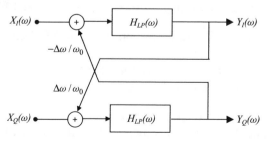

Figure 1.9 Implementation of a one-pole complex polyphase filter

For instance, if the quadrature branches have either a gain asymmetry of about 0.1 dB, or a phase asymmetry of about 0.7 degrees, the worst-case floor of the composite complex filter can be expected to be around 38 dB below band-pass.

1.2.3.3 VLIF Outlook

Since the complex polyphase filter can be entirely implemented by mathematical DSP computations, rather than with analog hardware, then it is possible to use the very same DCR architecture of Figure 1.7, where the only change required is the frequency of the LO. Thus, one can find many ICs implementing both DCR and VLIF on the same hardware. The VLIF advantages are many, but there are disadvantages:

THE BAD NEWS

- Low-cost VLIF implementations employ hardware-based polyphase filters, yielding intrinsic low-spec receivers with RF figures of the order of less than 50 dB, since the ultimate stop-band rejection of the complex polyphase filter is limited by the phase and amplitude balance accuracy between the I and Q branches.
- I and Q baseband channels must be processed at once as a complex signal.
- Since the IF frequency is greater than zero, the analog to digital (A/D) conversion as well DSP computations must be carried out at a speed faster than in DCR.

THE GOOD NEWS

- The architecture lends itself to easy integration just as the DCR.
- In applications where the cost/complexity is less critical, the polyphase filter can be implemented by DSP computations, yielding higher performance specs, due to perfect balance of filter branches.
- Impairments due to '$1/f$' noise, DC offset, self-mixing LO leakage etc. are greatly reduced.
- The front-end RF filter requirements are greatly reduced.

1.3 Transmitter Architectures

We arbitrarily divide the transmitter in three main functional blocks:

- Power amplifier (PA): all the amplifiers at final RF frequency with input > +10 dBm.
- Exciter: the amplifier chain whose output drives the PA.
- Backend: all the other circuits.

The critical parameter we focus on, in determining the performance of a transmitter, is the unwanted emitted RF power as a function of its distance form the central carrier, relative to the total transmitted RF power.

We always translate every disturbance (noise, synthesizer spectral purity, insta-bility, nonlinearity etc.) into an equivalent unwanted emitted RF power 'some-where' inside or outside the designated RF channel. By this we mean the total power obtained by integrating the unwanted spectrum over some given bandwidth centered at a given distance from the transmitted carrier.

The unwanted emitted RF power is measured at the antenna port in terms of parameters such as error vector magnitude (EVM), emission mask, adjacent channel power (ACPR) etc., discussed in detail in Chapter 3; and should be kept below a given level, which may be dictated by regulatory limitations, system performance requirements, or both.

The most important modern Tx architectures, which we focus on in this book, are the two-step conversion transmitter (TSCT) and the direct launch transmitter (DLT), which are useful for most broadband applications. These architectures make use of an IQ modulator line up, and are suitable for both constant-envelope modulation methods, such as BPSK; and variable-envelope modulation schemes, such as QAM.

The general RF signal $S(t)$, defined in Equation (1.1) is generated by applying the baseband signals $I(t) = s_r(t)$ and $Q(t) = s_i(t)$, with $s_r(t), s_i(t)$, defined in Equation (1.2), to an IQ modulator similar to that shown in Figure 1.10. The procedure is just the reverse of the one used in the receiver case to obtain $I(t)$ and $Q(t)$ from $S(t)$.

The baseband signals are handled, in most cases, by a DSP, which carries out signal processing, source coding, filtering, symbol mapping and shaping, inter-leaving etc. Simpler but widely used narrowband architectures using direct FM modulation and single-ended line-up will also be briefly treated. One of the most difficult subsystems to design and implement in modern digital transmitter design is the RF power amplifier. Discussion of trade-offs between efficiency, linear-ity, RF power output and power supply level (mainly an issue with low-voltage battery-operated portable equipment) is deferred to Chapter 3.

1.3.1 Two-Step Conversion Transmitter (TSCT)

The TSCT with IQ modulator is the right choice for delivering high-performance narrowband and broadband receiver specs. It is illustrated in Figure 1.10.

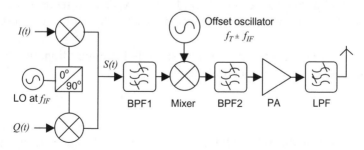

Figure 1.10   Two-step conversion transmitter architecture

1.3.1.1 TSCT Highlights

Looking at Figure 1.10, we see that the baseband signals $I(t)$ and $Q(t)$ generate, at the modulator output, a combined signal $S(t)$ at IF frequency of the form:

$$S(t) = I(t) \cos(\omega_{IF} t) - Q(t) \sin(\omega_{IF} t) \qquad (1.25)$$

Substituting in Equation (1.25)

$$I(t) = A(t) \cos(\varphi(t))$$
$$Q(t) = A(t) \sin(\varphi(t)) \qquad (1.26)$$

it is possible to verify that we obtain the general RF signal (1.1) at IF frequency.

The role of the filter BPF1 is to reject all undesired signals outside the required transmitter bandwidth at IF, such as LO harmonics, image noise etc. The mixer then simply shifts the IF signal to the final transmit frequency f_T. The filter BPF2 rejects unwanted image signal produced by the upmixing. The signal is then amplified by the power amplifier (PA) and transmitted through the LPF which suppresses PA harmonics. The matching network at the PA output is designed to optimize power output vs. linearity performance.

The TSCT approach has many advantages. Some of which are:

- The IQ modulator operates at a fixed frequency (IF), which makes it easy to get good and consistent phase and amplitude matching.
- The final carrier frequency is far from LO frequency. As we will see in Chapter 6, this frequency separation prevents the appearance of a serious parasitic effect known as 'VCO injection pulling'.

1.3.1.2 TSCT Outlook

The following is worth noting about TSCT architecture:

THE BAD NEWS

- The high-level performance is delivered at the expense of increased complexity, cost, current consumption, component count and physical size as compared to DLT.
- Moreover the TSCT architecture is less suitable for integration, due to the many connections to external lumped elements.

THE GOOD NEWS

- Due to the use of the IQ modulator at a single *IF* frequency, an outstanding overall performance can be achieved with relatively mild constraints on the

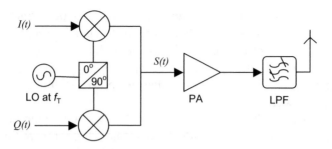

Figure 1.11   Direct launch transmitter architecture

active backend components, which reflects, among other things, in reliable and consistent operational figures.

1.3.2 Direct Launch Transmitter (DLT)

The DLT with IQ backend is suitable for delivering broadband data with mid-level performance specs. Most low-cost and low-power broadband applications, such as Bluetooth and WiFi, utilize DLT architectures, as they are well designed for CMOS SoC integration. The DLT architecture is described in Figure 1.11.

1.3.2.1 DLT Highlights

The baseband signals $I(t)$ and $Q(t)$ are converted directly to the final transmitter frequency f_T by the LO and summed together to get a combined signal $S(t)$ at the modulator output, in the form

$$S(t) = I(t)\cos(\omega_T t) - Q(t)\sin(\omega_T t) \qquad (1.27)$$

The signal $S(t)$ in Equation (1.27), is the general RF signal defined in Equation (1.1), and is generated in a way identical to one described to obtain the signal in Equation (1.25), except that the LO is at final frequency rather then at IF frequency.

As opposed to the TSCT case, the BPF at the modulator output is no longer required, as no image noise and no IF products are generated. The modulator output is delivered directly to the PA input. This signal is then amplified by the PA and transmitted through the LPF, which suppresses the PA harmonics. Here too, the PA matching network is designed to optimize power output vs. linearity. In DLT configuration, however, LO and PA operate at the same frequency. This fact makes the DLT transmitter prone to a major source of interference known as 'voltage-controlled oscillator (VCO) injection pulling', which may produce parasitic VCO modulation if the PA radiates onto the LO. 'Injection pulling' is analyzed in detail in Chapter 6.

What we can anticipate is that when PA and LO are close in frequency, the VCO tends to track the instantaneous frequency of the modulated carrier. This

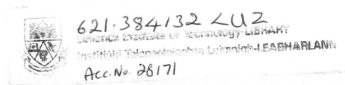

unwanted LO modulation degrades the total transmitter performance and causes the transmitter spectrum to widen. This phenomenon can be found in the literature under the name of 'spectrum regrowth' or 'remodulation'. In most cases, good shielding on the VCO and synthesizer area is required in order to reduce it. Additional filtering on the VCO and synthesizer power supplies and control lines may be needed as well.

Another issue with DLT, as opposed to TSCT, is that the IQ modulator no longer operates at fixed frequency, and needs to cover the whole transmitter band, which makes it difficult to obtain consistent phase and amplitude matching across all channels.

1.3.2.2 DLT Outlook

The following is worth noting about DLT architecture:

THE BAD NEWS

- Backend linearity, spectral purity and IQ balance constraints are tougher than in the TSCT case.
- Special care should be taken in shielding the VCO to avoid remodulation phenomena due to PA radiation. This architecture is more suitable for low-power applications.

THE GOOD NEWS

- The DLT architecture greatly simplifies the transmitter line-up, as can be appreciated by comparing Figure 1.11 to Figure 1.10. A single LO is required, and the IF chain is removed, leading to a major reduction in size and cost.

1.3.3 Direct FM Transmitter (DFMT)

The DFMT architecture is simpler than the DLT because only constant-envelope signals are employed. The DFMT architecture is described in Figure 1.12.

FM transceivers are widely used in military, public safety, and commercial equipment for both voice and data applications. Military and public safety applications consist, in most cases, of narrowband systems with channel width of

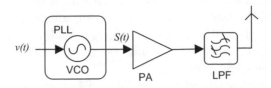

Figure 1.12   Direct FM transmitter

12.5 kHz and 25 kHz. In commercial applications, the most popular cellular world standard, the GSM (global system for mobile communications), uses an FM transceiver with Gaussian minimum shift keying (GMSK) modulation, transmitting over a 200 kHz channel. FM transceivers are attractive because they generate constant-envelope signals, which can be handled with a very simple and energy-efficient transceiver implementation. GSM handsets can be built using nonlinear RF PA technologies and simple receiver architectures that result in low-cost, low-current and small-size products.

For the sake of completeness, let us review briefly the fundamentals of FM modulation: an FM signal can be represented as a particular case of Equation (1.1) in the form

$$S(t) = A \cos \left[2\pi f_c t + 2\pi \Delta f \int^t v(\tau) \, d\tau \right], \quad |v(t)| \le 1 \tag{1.28}$$

where $v(t)$ is a normalized modulating signal. The expression in square brackets is the instantaneous signal phase, Δf is called 'the peak deviation', f_c is the carrier frequency and A is the signal amplitude, which in the FM case is constant.

The instantaneous angular frequency is the derivative of the instantaneous phase, thus $f_i = f_c + \Delta f v(t)$, and $|f_c - f_i| \le \Delta f$ by normalization of $v(t)$.

Constant-envelope digital modulations can be efficiently implemented using DFMT, which uses the baseband signal to directly frequency-modulate a VCO operating at the final transmitter frequency.

1.3.3.1 DFMT Highlights

The RF modulated signal $S(t)$ is obtained by directly modulating the VCO frequency with the baseband signal. The modulated signal is then passed to the PA and transmitted through the LPF to suppress PA harmonics. Since the signal has constant envelope, a nonlinear class-C PA can be used (Chapter 3). As compared to linear amplifiers a class-C PA has much better DC-to-RF efficiency, thus allowing for smaller sizes of power device and heat-sink, which reflects in lower costs. Class-C amplifiers also exhibit low-current consumption, which is a major feature for hand-held products.

The VCO in Figure 1.12 always operates in a closed loop as part of the phase-locked loop (PLL) section of the transceiver synthesizer. The PLL tends to keep the VCO frequency constant (Chapter 4), thus if the modulating signal falls inside the PLL bandwidth (which is about the reciprocal of synthesizer lock time), the modulation will be cancelled out by the PLL action. This effect is known as 'modulation track-out'.

In DFMT architectures, the PLL bandwidth is small (below 300 Hz), thus modulation track-out usually does not affect analog voice, which is within the 0.3–3 kHz band, or high-rate data signals. However, when low-speed data is employed (such as simultaneous sub-band non-return-to-zero (NRZ) control signals), unless the PLL bandwidth is made very small, a special arrangement known as 'dual port modulation' is required to avoid track-out.

Dual-port modulation is the preferred choice since, in practice, one cannot reduce the PLL bandwidth too much, otherwise:

- A very slow PLL will yield an unreasonably long synthesizer lock time.
- The beneficial effect known as 'microphonics track-out' will be lost. Microphonics track-out is the bright side of modulation track-out, namely the cancellation of the parasitic VCO modulation caused by physical vibrations, which usually occurs at frequencies below the PLL bandwidth (Chapter 6).

1.3.3.2 DFMT Outlook

The following is worth noting about DFMT architecture:

THE BAD NEWS

- DFMT is suitable for constant-envelope applications only.
- If the modulating signals contain low-frequency spectral components (below the reciprocal of synthesizer lock time), dual-port modulation will be required, which makes the implementation somewhat more complex and expensive.

THE GOOD NEWS

- The DFMT architecture greatly simplifies the transmitter line-up, and is well suited for low-cost, low-current voice-oriented applications. Mixer, IF chain or quadrature (IQ) modulation are not required, leading to a major reduction in size and cost.

1.3.4 Dual-Port Direct FM Transmitter (DDFMT)

As we have seen in section 1.3.3 the DFMT architecture is less suitable for use when the baseband signal contains spectral components in the vicinity of DC at very low frequencies. The PLL response for baseband signals modulating the VCO is of high-pass nature, while a low-pass response can be achieved by modulating the PLL frequency divider or the reference oscillator (Chapter 4). To achieve a wideband modulated RF signal the baseband signal needs to modulate the VCO and the frequency divider simultaneously. The DDFMT architecture is described in Figure 1.13.

1.3.4.1 DDFMT Highlights

The baseband signal $v(t)$ is split into two paths, one is directed to the VCO and the other to the divider. The low-frequency baseband components (below the

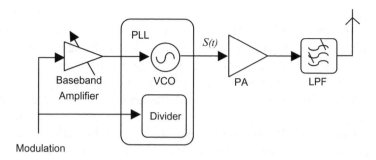

Figure 1.13   Dual-port FM transmitter

loop bandwidth) modulate the divider or the reference oscillator, while the high-frequency components (above the loop bandwidth) modulate the VCO. As a result, the composite PLL response is all-pass and includes all the baseband components including DC (corresponding to a frequency shift of the carrier signal). This configuration is frequently employed for NRZ, or multilevel baseband data. The purpose of the gain-adjustable baseband amplifier is to balance the peak deviation of the two paths, in order to avoid distortion of the combined signal.

1.3.4.2 DDFMT Outlook

The following is worth noting about DDFMT architecture:

THE BAD NEWS

• DDFMT is suitable for constant-envelope applications only.

THE GOOD NEWS

• The architecture greatly simplifies transmitter line-up. Mixer and IF chain are not required, leading to a major reduction in size and cost.

1.4 Transceiver Architectures

Modern wireless systems are designed to support multiple users coexisting and working simultaneously in a variety of crowded and often wild electromagnetic environments, and may successfully operate only with associated subscriber equipment carefully tailored to specific requirements.

A transceiver, designed with improper architecture, may either work erratically under particular scenarios or produce dramatic interferences capable of paralyzing its own home network or severely impairing neighboring systems. Indeed, to prevent such 'system killers', cellular and other operators impose severe certification procedures before they allow manufacturers to introduce new subscriber equipment.

In this section we describe transceiver architectures for use with widely accepted access methods, such as FDMA (frequency division multiple access), TDMA (time division multiple access), and CDMA (code division multiple access). A thorough discussion of access methods is beyond the scope of this book and we refer the reader to the literature for more insight, see Dixon (1994).

Here we only mention that, in principle, the access methods mentioned above determine the way by which multiple users may physically share (access) two common sets of paired RF channels, one assigned to uplink (the subscriber transmit channel), and the other to downlink (the subscriber receive channel). In most access methods utilizing paired channels, the uplink channel is the one lower in frequency.

We also analyze an architecture suitable for TDD (time division duplex) operation, an emerging RF transmission technology of primary importance in fourth-generation broadband systems. TDD is a single-channel access method that allows each user to optimally utilize the assigned bandwidth by dynamically partitioning it between transmitted and received data, and is especially effective in systems whose uplink and downlink channels have asymmetric or variable data rates. Since the uplink and downlink channels are at the same frequency, TDD is also very suitable for the implementation of 'smart antenna' technologies, such as MIMO (multi-input multi-output) and adaptive-phased arrays (Gross, 2005). This is because at a fixed channel frequency, the short-term RF path characteristics are relatively stable, and thus one may switch between transmit and receive mode while having to re-optimize the antenna array only once in a while, saving both overhead time and substantial computing power.

1.4.1 Full-Duplex CDMA/FDMA Architectures

Commercial CDMA and FDMA transceivers usually work in full-duplex mode. In other words, receiver and transmitter operate simultaneously in different and widely spaced frequency bands through a common antenna.

The wide Tx-Rx separation, enables the use of isolating devices known as 'duplexers'. A typical implementation is shown in Figure 1.14.

A duplexer is in essence a couple of bandpass filters, one around the transmit band and the other around the receive band. Since the receive filter rejects the transmit frequency, then it has nearly purely imaginary impedance in the transmit frequency band. This imaginary impedance may be easily 'rotated' (on a Smith chart) by use of a coaxial cable or a microstrip line of proper length, so as to appear as an open circuit at the transmitter output (Collin, 1966).

It follows that the receiver port does not load the power amplifier, and there is very little transmit power coupled into the receiver input. Thus, receiver damage as well as desense (Chapter 2) are prevented. An identical arrangement rotates the impedance of the transmit filter and prevents the transmitter output from loading the receiver input. As a result of the above, a duplex radio may transmit and receive simultaneously and continuously through one common antenna.

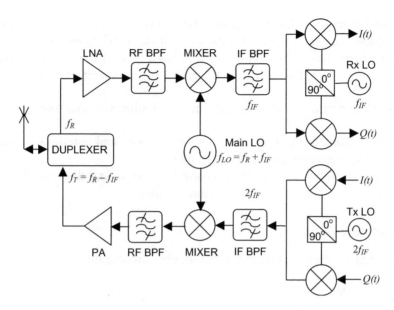

Figure 1.14   Full-duplex architecture

In some applications, the receiver is isolated from the transmitter by the use of two separated Tx and Rx antennas. However, this usually requires substantial (vertical) physical separation between the antennas and is suitable for fixed applications only.

Indeed, the isolation achievable by two antennas located on a portable device is as low as 15 dB, which, as we will see, is absolutely inadequate for full-duplex operation.

Full-duplex architectures where introduced in the late 1980s and early 1990s in analog cellular systems such as AMPS and NAMPS, and are still employed in modern CDMA and WCDMA digital cellular systems.

1.4.1.1 Full-Duplex Highlights

Transmit and receive signals can be applied simultaneously at the duplexer input, provided that the frequency separation between the two is wide enough. For instance, in CDMA systems operating in the 800 MHz band, the Tx–Rx frequency separation is 45 MHz, while in systems working in the 1800–1900 MHz band, the separation is 90 MHz.

At the transmitter side, the duplexer has several functions:

- Acts as a harmonic filter to 'clean up' the transmitted signal due to out-of-band spurs.
- Prevents far-out transmitter phase noise from entering the receiver. Since the transmitter phase noise (Chapter 5) reaches a far-out floor, whose power is

much higher then the lowest signal to be received, without the blocking action of the duplexer, strong receiver desense will occur (Chapter 2).

• The imaginary output impedance of the transmitter filter at the receiver frequency, can be 'rotated' to an open circuit preventing receiver loading.

The receiver portion of the duplexer filter is multifunctional too:

• Performs the functions of the receiver 'front filter', and determines intermodulation and spurious specifications (Chapter 2).
• Has imaginary impedance at the transmit frequency, and thus may be 'rotated' to an open circuit, preventing transmitter loading.

1.4.1.2 Full-Duplex Outlook

The following is worth noting about full-duplex architectures:

THE BAD NEWS

• The duplexer, in most cases, is a bulky and high-cost unit, which becomes bigger in size and more complex as transmitter power increases or Tx–Rx separation decreases.
• The simultaneous receiver and transmitter operation creates a considerable number of extra spurious signals not present in simplex radios and may considerably deteriorate receiver performance (Chapter 2).

THE GOOD NEWS

• The ability to transmit and receive signals simultaneously is a major advantage in systems working on paired channels, whenever the time overhead required to switch from transmit to receive state (and vice versa) cannot be tolerated.

1.4.2 Half-Duplex/TDMA Architectures

Commercial TDMA systems work in half-duplex mode on paired channels. In other words, receiver and transmitter do not operate simultaneously, however, the uplink and downlink frequencies are different, with large Tx–Rx frequency separation. The reason for the large frequency separation is that, although the remote units work in half-duplex mode, the base station must still have full-duplex capability in order to be able to effectively control the subscribers. In turn, full-duplex capability implies the use of a duplexer, which dictates large Tx–Rx frequency separation. Figure 1.15 describes a half-duplex transceiver based on the full-duplex concept. Since the Tx–Rx transition does not require a synthesizer jump, this architecture is capable of working in fast TDMA mode. If the TDMA system works in slow mode, say, with a slot time of several milliseconds, then a

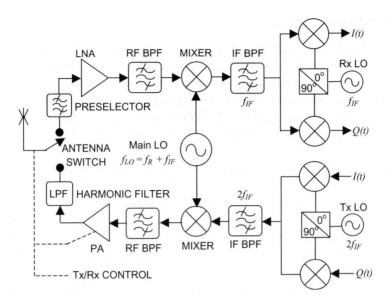

Figure 1.15   Fast TDMA half-duplex architecture

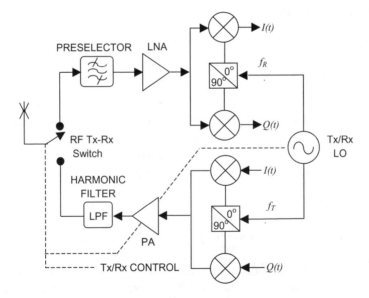

Figure 1.16   Slow TDMA half-duplex architecture

much simpler approach may be taken, where a common fast synthesizer (usually of the fractional-N type described in Chapter 4), which continuously jumps between Tx and Rx frequency, is used for both the transmit and the receive path of a DCR–DLT combination as shown in Figure 1.16. Modern systems such GSM and iDEN make use of fast TDMA and slow TDMA architectures, respectively.

1.4.2.1 Half-Duplex Highlights

In the fast TDMA transceiver of Figure 1.15, the duplexer is replaced by an RF switch, usually referred to as the 'antenna switch', which periodically connects either the transmitter or the receiver to the common antenna, according to the Tx–Rx time-slot allocation. Since no duplexer protection exists, a preselector filter must be added (see section 1.2.1.1) to reject receive spurs (see Chapter 2), and a low-pass filter, usually referred to as the 'harmonic filter', to clean up PA harmonics.

A half-duplex radio cannot transmit and receive simultaneously. The transmitter must be shut down before disconnecting it from the antenna, and operated again only after reconnection has been completed. The dotted line in Figure 1.15 stresses the fact that, although the synthesizer is static, the antenna switch and the PA must operate in synchrony. The antenna switch must have just enough isolation to prevent damage to the receiver, but there are no desense issues, as the receiver is not operational during transmission.

In the slow TDMA transceiver of Figure 1.16, the Tx–Rx synthesizer too must be synchronized with the antenna switch and the PA.

1.4.2.2 Half-Duplex Outlook

The following is worth noting about half-duplex architectures:

THE BAD NEWS

- The intermittent working mode implies some undesired overhead due to the introduction of guard times for both transmitter and receiver, including the change of state of the antenna switch, the time for the transmit power to build up/decay, the time for the receiver to synchronize, and, in slow TDMA architecture, the time for the synthesizer to lock up.

THE GOOD NEWS

- The RF switch is simpler, less bulky, and in most cases, has smaller losses than a duplexer.
- Moreover, TDMA operation allows for some freedom for dynamic bandwidth allocation, since one may decide 'on the fly' how many slots to assign to each user.

1.4.3 Simplex/TDD Architectures

Simplex radios are the most ancient type of transceivers, and, as suggested by their name, operate on the same channel frequency alternating transmit and receive mode, in a 'single-channel TDMA' fashion. They have been disregarded for a long time, while second and third cellular generations spread, and they survived mainly in military or marine applications where a central base site cannot be relied upon.

With the advent of broadband wireless systems, simplex radios working in TDD mode are reviving, due to their flexibility in bandwidth allocation and their capability to work efficiently without a supporting wireless infrastructure, which is typically missing when operating in unlicensed bands.

The typical TDD architecture shown in Figure 1.17 is identical to the fast TDMA half-duplex transceiver of Figure 1.15, except that uplink and downlink frequencies have the identical value f_0. This is easily obtained by using a common synthesizer for both transmit and receive paths. The synthesizer works at a fixed frequency, so the radio may alternate between receive and transmit mode with no lock time overhead.

One may object that if we take the slow TDMA radio of Figure 1.16, and set its synthesizer to work at the fixed frequency f_0, both radios will operate identically, while the slow TDMA architecture is much simpler. This is true, but, unfortunately, using such a DCR–DLT combination in a TDD system, will likely introduce a system killer that will jeopardize the wireless network in its neighborhood. The reason for such a disaster lies in the fact that, while in receive mode, the LO of the DCR is exactly at channel frequency. This LO carrier will either leak to the antenna through the backward Rx path, or, even worse, radiate directly form the radio body, and, since TDD systems work in 'carrier-sense' mode, the channel will be constantly held busy, preventing nearby radios from transmitting. Horror stories of this kind occurred in the past, when operating

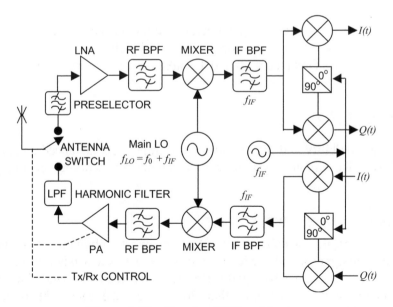

Figure 1.17 Simplex TDD architecture

analog full-duplex radios on half-duplex systems. This will not happen with the simplex TDD architecture of Figure 1.17, since there are no oscillators at the channel frequency.

So we see that while two radios fulfill the same functional tasks, one of them is not suitable for operating in the system. Simplex TDD architectures are in use with WLAN (wireless local area network) systems and last-mile WiMax applications.

1.4.3.1 Simplex Highlights

As pointed out before, the fact that one works on a single channel leaves to the user the decision of how to partition the time between transmit and receive mode. In asymmetric channels this is a nice feature, for if one tries to work in ADSL (asymmetric digital subscriber loop) mode in a full-duplex or half-duplex system, the uplink bandwidth will not be fully utilized, while the downlink will be choked. Also, as stressed before, working on a single channel significantly reduces the overhead time required to 'learn' the channel characteristics, since the learning process does not have to be repeated at each Tx–Rx transition. One example of this type is the WLAN 802.11g, which operates in the 2.4 GHz band in TDD over 20 MHz bandwidth.

1.4.3.2 Simplex Outlook

The following is worth noting about simplex architectures:

THE BAD NEWS

- Receiver and transmitter cannot operate simultaneously. The single channel approach makes the system prone to more disturbances that in the case of paired channels. Great care should be taken in the design phase to prevent system-killer effects.

THE GOOD NEWS

- The great flexibility and the opportunity to implement advanced antenna technologies make simplex systems a great choice for working on unregulated bands.

1.4.4 Ultra-Wideband (UWB) Systems

The definition of ultra-wideband (UWB) is somewhat ambiguous. It is customary to designate a signal as ultra-wideband if either it has a bandwidth greater that 500 MHz, or a fractional bandwidth (the ratio between the bandwidth and the carrier center frequency) of more than 20%, where the bandwidth is measured between -10 dB points.

UWB systems are intended to transfer data at the rate of several hundreds Mbps over short distances, up to few meters, while simultaneously coexisting with other signals of different types in the same spectral bandwidth.

The UWB idea has much in common with the CDMA and OFDMA (orthogonal frequency division multiple access) approaches: as the transmitted signal is more and more spread out in frequency, the RF interfering power it generates within narrowband channels or within other UWB signals coded differently, will be less and less noticeable.

There are two widely accepted (and competing) technologies for implementing UWB transceivers: direct-sequence-coded sub-nanosecond impulses (DS-UWB) and multi-band orthogonal frequency-division multiplexing (MB-OFDM). Each technology has its peculiar advantages and drawbacks, and, at this point in time, it is not clear if one of the two will prevail.

UWB has found other uses outside the communication arena, including specialized applications such as ground-penetrating radar, through-wall imaging, medical imaging and location finding. The reason why UWB is so suited for radar applications lies in its wide bandwidth. In fact, given a bandwidth B, and denoting by c the speed of light, the smallest physical dimension that can be measured by means of an RF signal is roughly c/B, thus, with a bandwidth larger than 500 MHz, one may detect objects smaller than 50 cm.

In this book we will not discuss UWB beyond this point, as it substantially differs from most other communication systems from the standpoint of both technology and applications, and deserves a dedicated textbook. A comprehensive overview of UWB communication systems can be found in Fontana (2004), Runkle et al. (2003), and Saberinia and Tewfik (2003).

2

Receiving Systems

In this chapter, we define and explain in detail all the various receiver specifications, and we show how to achieve, compute and measure them.

As far as we are concerned at this point, receivers are built of 'black boxes' and we are not concerned with what is inside them. We have an input port at the antenna, and all we care about is the signal quality at a virtual output port at the detector input. Post-detector functionality is irrelevant at this stage.

With the practical side in mind, we first define the useful definitions and results in a format ready for use in design and analysis, and then the essential detail follows. Square brackets [ ] denote dimension, and logarithmic representations such as dB, dBc, dBm etc., are used *only* when explicitly specified.

2.1 Sensitivity

The receiver (Rx) 'sensitivity', which we denote by *Sens*, is the smallest RF power at the antenna input and at operating frequency (on-channel, center-frequency), which will produce a 'useful' signal at the input of the detector when no other RF signals are interfering at the antenna port. By 'useful' we mean a signal that yields some predefined signal-power-to-noise-power ratio into the detector, which we denote by *SNRd*. Thus, assuming a center-frequency power gain G from antenna to detector, and a noise power Nd at detector input, the sensitivity is the smallest RF signal power satisfying:

$$\frac{Sens \cdot G}{Nd} = SNRd \qquad (2.1)$$

The noise power Nd is the result of integrating the thermal noise generated in all the circuits preceding the detector over the bandwidth seen at detector input. It turns out that an *SNRd* of 10 is a 'magic number', corresponding roughly to the level at which the receiver output starts becoming usable in the majority of common applications. Therefore, in all that follows, we choose, as a rule of thumb, to define the receiver sensitivity as the RF signal level at the antenna port

Wireless Transceiver Design   Ariel Luzzatto and Gadi Shirazi
© 2007 John Wiley & Sons, Ltd

that will yield about 10 dB *SNRd*. To this RF level corresponds a signal-power-to-noise-power ratio at the antenna input *SNRi*. Of course, the same *SNRd* figure for two different receivers may correspond to two different values of *SNRi*.

2.1.1 Computation of Sensitivity

The main parameters limiting the receiver sensitivity are:

- Noise figure: $NF = 10\,log\,(SNRi/SNRd)$ [dB].
- RF channel bandwidth B [Hz] seen at detector input.
- The *SNRd* required for sensitivity performance.

Once the above parameters are known, the sensitivity in dBm is given by:

$$Sens|_{dBm} = -174 + 10\,log\,(B) + NF + 10\,log\,(SNRd)\ [dBm] \qquad (2.2)$$

The following computation proves Equation (2.2): the thermal noise power generated at antenna port within a given bandwidth B (Schwartz, 1990) is KTB [watt], where $K = 1.38 \times 10^{-23}$ is the Boltzmann constant in MKS units, and $T = 300$ is the room temperature in degrees Kelvin. At the antenna we have $SNRi = Sens/KTB$. Thus $NF = 10\,log\,(Sens/KTB/SNRd)$, from which Equation (2.2) is obtained by computing *Sens* and then setting

$$Sens|_{dBm} = 10\,log\,(Sens/10^{-3})$$

One particular issue with very-low IF (VLIF) receivers is worth noting: the thermal noise generated at the antenna port has extremely wide bandwidth, but usually receivers working with high-frequency IF (superheterodyne) include a front-end filter that strongly attenuates signals at image frequency (see section 2.6). The side benefit is that virtually no thermal noise is being generated at image frequency, because the out-of-band filter impedance is nearly purely imaginary. In contrast, for practical reasons, VLIF receivers have a front filter much wider than the IF frequency, thus the noise power at image frequency reaches the detector, which results in worsening the overall receiver noise figure (*NF*).

Example: Consider a GSM cellphone. We assume: $NF \approx 8$ dB and bandwidth $BW \approx 140$ kHz (due to guard-band, *BW* is somewhat smaller than the channel spacing, which is 200 kHz). With $SNRd \approx 10$, Equation (2.2) yields $Sens \approx -104.5$ dBm.

2.1.1.1 Computation of SHR Sensitivity

Let us recall first how to compute the *NF* value seen at the input of a cascaded two-stage system. For a power gain G and an *NF* in dB, the corresponding

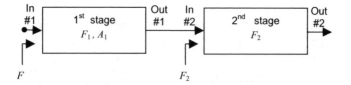

Figure 2.1 Computation of input noise figure for cascaded stages

numeric values are denoted respectively by $A = 10^{G/10}$ and $F = 10^{NF/10}$. With reference to Figure 2.1, let

$$F_2 = SNR|_{\text{input \#2}}/SNR|_{\text{output \#2}} : \text{noise figure of stage \#2}$$

$$F_1 = SNR|_{\text{input \#1}}/SNR|_{\text{output \#1}} : \text{noise figure of stage \#1}$$

The overall two-stage noise figure $F = SNR|_{\text{input \#1}}/SNR|_{\text{output \#2}}$ has the numeric value:

$$F = F_1 + \frac{F_2 - 1}{A_1} \tag{2.3}$$

Also recall that the *NF* of a passive element is equal to its attenuation.

Now, let us compute the sensitivity for the SHR architecture of Figure 1.1, using the subsystem values given in Table 1.1 and repeated at the top of Figure 2.2.

The *NF* of the backend used in the receiver of Figure 2.2 is the ratio between the *SNR* at backend input and *SNRd* at detector input (which may be either inside the backend chip itself, or implemented somewhere else, for instance in a DSP external to the backend chip).

Now, the *NF* at receiver input can be computed by using Equation (2.3) recursively, starting from the input of the backend, and going back step by step towards the input of the first band-pass filter (BPF). The computed *NF* value (in dB) for each backwards step is shown at the bottom of Figure 2.2. The overall *NF*, 'looking' into each inter-stage point, is the ratio between the *SNR* at that point and *SNRd* (at detector input). Finally, the computed input noise figure, *NFi*, yields the ratio between *SNRi* (the *SNR* at receiver input) and *SNRd*. Using our rule of thumb for sensitivity, we require *SNRd* = 10.

Then, with the help of the computed *NF* and substituting the IF bandwidth for *B* in Equation (2.2), we are able to compute the power *Sens* of the RF signal required for sensitivity performance at the input of each stage in the receiver chain, with the previous stages taken out. The overall receiver sensitivity (in dBm) and *NF* (in dB), as well as the interim sensitivity and *NF* values obtained at the input of each stage, are also shown at the bottom of Figure 2.2. The interim computation is useful during the design process, since it enables us to compare the measured interim sensitivity to the theoretical one, confirming the accuracy of the design before we continue to the next stage. Note that, for computation

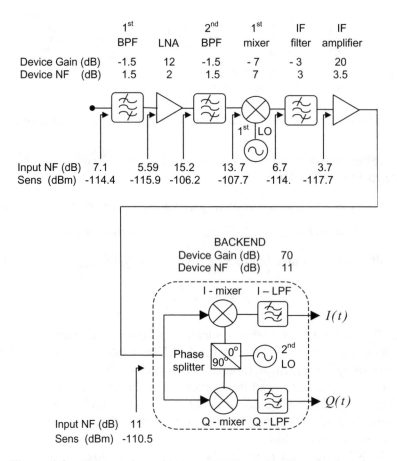

	1st BPF	LNA	2nd BPF	1st mixer	IF filter	IF amplifier
Device Gain (dB)	-1.5	12	-1.5	- 7	- 3	20
Device NF (dB)	1.5	2	1.5	7	3	3.5

Input NF (dB)	7.1	5.59	15.2	13. 7	6.7	3.7
Sens (dBm)	-114.4	-115.9	-106.2	-107.7	-114.	-117.7

BACKEND

Device Gain (dB)	70
Device NF (dB)	11

I - mixer   I – LPF

$I(t)$

Phase splitter   0°   90°   2nd LO

$Q(t)$

Q - mixer   Q - LPF

Input NF (dB)	11
Sens (dBm)	-110.5

Figure 2.2   Computation of interim SHR sensitivity and noise figure

purposes, the backend pass-band is the same as the IF pass-band, since both I and Q channels are 9 kHz wide, yielding an equivalent noise bandwidth of 18 kHz.

NOTE: It may be that both sensitivity and noise figure are worse at the input than near the backend. This fact *does not imply* that the front stages are badly designed, but is the outcome of the fact that sensitivity, in spite of being an easy spec to obtain, always comes at the expense of other crucial figures. In actual designs, sensitivity is frequently sacrificed along the receiver RF chain, in favor of intermodulation, spurious response etc.

2.1.1.2 Computation of DCR Sensitivity

Figure 2.3 shows the computation of the interim sensitivity and *NF* of the DCR architecture of Figure 1.7 using the subsystem values given in Table 1.2 and repeated at top of Figure 2.3. The definitions of section 2.1.1.1 and section 2.1.1 apply.

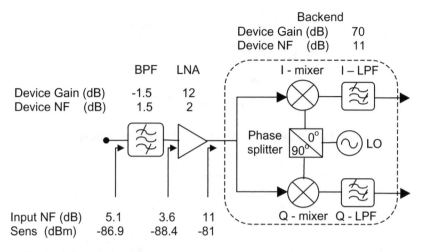

Figure 2.3 Computation of interim DCR noise figure and sensitivity

Again we require *SNRd* = 10, but here, we substitute for *B* in Equation (2.2) the given IF bandwidth of 16 MHz.

2.1.1.3 Computation of VLIF Sensitivity

The computation for VLIF is essentially identical to the DCR case, since the line-up from antenna to backend input is similar, and the architectural differences in the backend itself do not affect it. This will not be true with other specifications such as selectivity, image rejection etc. as we will see later.

2.1.2 Measurement of Sensitivity

A laboratory setup for measuring sensitivity is described in Figures 2.4 and 2.5. There are no special requirements on the RF signal generator performance (sensitivity is an easy measurement in terms of equipment complexity requirements). The procedure is as follows:

- Set the RF generator level to minimum and adjust the frequency to the desired channel, then connect the generator output to the Rx antenna port.
- Increase the RF level until the Rx output reaches the sensitivity performance, according to whatever figure is considered as 'usable'. For instance, this may be a 10^{-2} BER value for digital equipment, a 12 dB *SINAD* value for analog FM, or even the direct measurement of an *SNRd* value of 10 (or other required figure) if the detector input is accessible.

Figure 2.4 shows the setup for the measurement using digital data modulation. Here the RF signal generator is modulated using a data stream generated by an

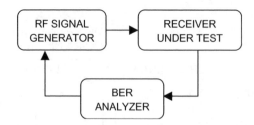

Figure 2.4   Sensitivity measurement with digital modulation

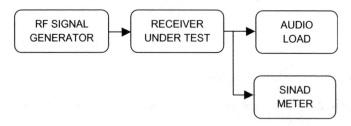

Figure 2.5   Sensitivity measurement with analog modulation

instrument called a 'BER analyzer'. The BER analyzer collects the data stream recovered by the Rx detector output, compares it with the data stream fed to the RF generator and computes the BER.

In other settings, for instance when using an IP protocol, it is customary to check the packet error rate (PER). For small values of PER, the BER can be roughly estimated by dividing the PER by the packet size, since it can be assumed that, on the average, there is not more than one erroneous bit in each packet. In this case it is a good practice to use the shortest packet size available in order to reduce the chance of having more than one wrong bit per packet, which may lead to inconsistent measurements.

Figure 2.5 shows the setup for the measurement using analog FM modulation. Here the RF signal generator is modulated using a 1000 Hz sinusoidal audio signal. The *SINAD* meter places a notch filter on 1000 Hz at the receiver audio output port, and computes the ratio between the total signal power (signal + noise + distortion) and the residual signal power (noise + distortion) after notch-out of the modulating sinusoid.

NOTE: Most receiver tests begin with a sensitivity measurement.

2.2 Co-Channel Rejection

Denote by $Si = 2Sens$ the power of an on-channel RF signal, which is twice the sensitivity power. Clearly, with Si applied at the antenna port, the Rx output performance will be better than the performance seen at the sensitivity level.

Denote by *Scc* the power of the smallest on-channel interfering RF signal that when applied to the antenna port simultaneously with *Si*, will cause the receiver output to worsen to the performance obtained at the sensitivity level.

The co-channel rejection, which we denote by *CCR*, is the ratio, in dB, between the power values *Scc* and *Sens*, namely:

$$CCR = 10\,log\,(Scc/Sens) \tag{2.4}$$

2.2.1 Computation of Co-Channel Rejection

The computation of *CCR* is identical for all the architectures described in section 1.2. In fact, the *CCR* is a measure of the *SNRd* at sensitivity level, and Equation (2.4) is roughly equivalent to Equation (2.5) below:

$$CCR = -10\,log\,(SNRd)\,[dB] \tag{2.5}$$

The following proves Equation (2.5): with *G* and *Nd* as in Equation (2.1), and with the interferer *Scc* and signal *Si = 2Sens* applied to the antenna input, the receiver performs as at sensitivity level, then *SNRd* is roughly the same as at sensitivity level (roughly because the co-channel interference does not behave exactly like thermal noise). It follows from Equation (2.1) that

$$\frac{1}{SNRd} = \frac{Nd}{Sens \cdot G} \approx \frac{Scc \cdot G + Nd}{Si \cdot G} \tag{2.6}$$

substituting *Si = 2Sens* into Equation (2.6) we get

$$\frac{Scc}{Sens} + \frac{Nd}{Sens \cdot G} \approx \frac{2}{SNRd} \tag{2.7}$$

Thus, again using (2.1) in (2.7), we finally get Equation (2.8), and taking its logarithm completes the proof.

$$\frac{Scc}{Sens} \approx \frac{1}{SNRd} \tag{2.8}$$

Example: Assuming our 'magic number' of 10 for *SNRd*, the measured *CCR* predicted by Equation (2.5) is −10 dB. As an example, the actual figure measured in most standard two-way FM radios is around −7 dB to −10 dB.

2.2.2 Measurement of Co-Channel Rejection

Figure 2.6 shows the measurement of *CCR* using digital modulation. The measurement using analog modulation is readily inferred from Figure 2.5. Here too there are no special requirements on the RF signal generator's performance. The

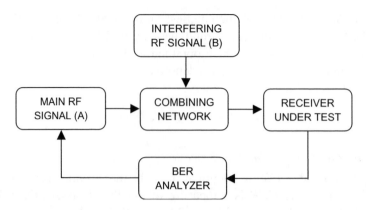

Figure 2.6   CCR measurement with digital modulation

CCR is measured by combining two RF signal generators, both tuned at center frequency.

First the interfering generator is shut off, and the level of the main generator (A) is increased until the output from the receiver fits the sensitivity performance (note that due to the attenuation of the combining network, the main signal level will be higher than the sensitivity level).

Then, the level of the main generator is increased by 3 dB, causing the output performance (BER or SINAD) to be better than the value at sensitivity. Now, the interfering RF signal generator (B) is modulated either with an analog or a digital signal, and its level is increased until the receiver output performance deteriorates to the sensitivity value.

The *CCR* is the ratio between the power level of RF signal B and the power of RF signal A at this stage (we implicitly assume that the combining network attenuation is the same for both generators A and B).

Care should be taken in the following:

- The modulation of signal B should be chosen so that it cannot be confused with the modulation of signal A. For instance, in analog FM measurements, if signal A is modulated with a 1000 Hz sinusoid, then signal B will be modulated with a 400 Hz sinusoid.
- The modulation level of signal B should be such that it does not cause the RF signal to exceed the channel bandwidth, otherwise part of the interfering power will be lost, and the measurement will be incorrect.

2.3 Selectivity

Consider again $Si = 2Sens$ as defined for *CCR*. Denote the on-channel frequency by f_R, and the spacing between adjacent channels by Δf. Denote by *Sadj* the smallest RF signal power at $f_R \pm \Delta f$ (one channel higher or lower than the operating channel) that, when applied to the antenna port simultaneously with

Si, will cause the receiver output to deteriorate to the performance obtained at sensitivity level.

The Rx 'selectivity', which we denote by *Sel*, is the ratio, in dB between the power values *Sadj* and *Sens*, namely

$$Sel = 10\log(Sadj/Sens) \qquad (2.9)$$

As for *CCR*, the interference generated doubles the noise power at the detector input, returning the *SNRd* to the same value as at sensitivity level. As opposed to the *CCR* case, however, the above interference is due to a number of different mechanisms adding up at once.

We anticipate that, as far as the measurement of selectivity is concerned, the phenomena listed below are indistinguishable, and all add up to the same effect of worsening the effective sensitivity measured in the presence of a signal at an adjacent channel frequency:

- Rx synthesizer phase noise *L* [dBc/Hz] at $f_{LO} \pm \Delta f$.
- RF channel bandwidth *B* [Hz] at detector input.
- Synthesizer spurs *SPRS* [dBc] at $f_{LO} \pm \Delta f$.
- Baseband/IF filter rejection *IFR* [dB] at $f_{IF} \pm \Delta f$.

It is worth pointing out, and we will show later on, that the synthesizer phase noise and spurs, and *not* the IF filter shape and bandwidth, are usually the limiting factors in this specification (except possibly in VLIF receivers). This is due to the fact that, even if the adjacent channel interferer itself is a perfectly clean 'delta' function, the synthesizer phase noise and spurs phase modulate it within the mixer (recall from Chapter 1 that all the mixer is doing is summing/subtracting phases), thus causing the (mixed) spectrum to acquire the very same shape of the LO, and to spread out into the operating channel.

2.3.1 Computation of Selectivity

In the following, Δf denotes an arbitrary frequency offset from the LO carrier. The cumulative effect of the phenomena triggered by the presence of the adjacent-channel interfering power *Sadj* determines the selectivity *Sel* according to

$$Sel = -10\log[10^{-IFR/10} + 10^{SPRS/10} + 10^{SBN/10}] + CCR \ [dB] \qquad (2.10)$$

where the parameters determining Equation (2.10) are as follows:

- $L(\Delta f)$ [dBc/Hz], the SSB (single-sideband) phase noise, is a property of the VCO/synthesizer combination (Chapter 5). It decreases when increasing the absolute frequency offset $|\Delta f|$ away from the carrier frequency and increases when increasing the LO frequency f_{LO}. In essence this is a parasitic phase modulation of the LO caused by thermal noise. As pointed out before, by the mixer action, the LO phase modulation is exactly transferred to the interferer, which spreads out to the operating channel. The selectivity is related to the close-in (near f_{LO}) value of the SSB noise power. Denoting the LO power by S_{LO}, we have, at $f_{LO} + \Delta f$

$$L = 10 \log(PHN_{LO}/S_{LO}) \ [dBc/Hz] \tag{2.11}$$

where we denote the synthesizer phase noise density at $f_{LO} + \Delta f$ by PHN_{LO} [watt/Hz], and dBc reads as 'dB below carrier'. We anticipate here that usually, for a given LO frequency f_{LO} and for Δf in the range concerning selectivity, $L(f_{LO}, \Delta f)$ behaves according to the simplified Leeson's equation

$$L(f_{LO}, \Delta f) \approx A(f_{LO}) - 20 \log_{10}(|\Delta f|) \ [dBc/Hz] \tag{2.12}$$

For f_{LO} fixed, Equation (2.12) shows that $L(\Delta f)$ drops by 6 dB each time the frequency offset away from the LO carrier is doubled. For very large offsets away from f_{LO} the simplified Equation (2.12) is no longer valid, and $L(\Delta f)$ reaches an ultimate floor. An in-depth analysis and insight into oscillator noise is given in Chapter 5.

Note that [dBc/Hz] in Equation (2.11) is a convention rather than a dimension. It means that PHN_{LO} at frequency $f_{LO} + \Delta f$ is

$$PHN_{LO} = 10^{\frac{L}{10}} S_{LO} \ [watt/Hz] \tag{2.13}$$

Denote by SBN (sideband noise) the value in dB obtained by integrating PHN_{LO}/S_{LO} over the channel bandwidth B. Denote by $Sadj_m$ the signal power resulting from mixing the adjacent channel interferer $Sadj$ with the LO, and by $PHNadj_m$ the phase noise density of $Sadj_m$ at some offset Δf from its center frequency. Since $Sadj_m$ acquires the very same spectral shape of the LO, then the relative phase noise density $PHNadj_m/Sadj_m$ as a function of the offset from $Sadj_m$, is identical to the relative LO phase noise density PHN_{LO}/S_{LO} as a function of the offset from LO carrier frequency f_{LO}, which produces at the detector input a relative power equal to SBN. If we consider PHN_{LO} as being nearly flat close to $f_{LO} + \Delta f$, then

$$SBN \approx 10 \log(PHNadj_m/Sadj_m \cdot B)$$
$$= 10 \log(PHN_{LO}/S_{LO} \cdot B) \tag{2.14}$$

and from Equation (2.11) it follows that

$$SBN \approx L + 10 \log(B) \ [dBc] \tag{2.15}$$

- The synthesizer spurs are 'delta-type' disturbances, and mostly depend on synthesizer hopping speed and loop filter construction. The most infamous are the 'reference spurs'. As pointed out before, the spurs modulate the interferer, which spreads out to the operating channel. Assuming for the sake of simplicity that we have just one dominant spur of power Sp at the adjacent channel, the relative noise produced in the detector input is

$$SPRS = 10 \log (Sp/Sadj) \ [dBc] \tag{2.16}$$

- The baseband/IF filter rejection IFR [dB] is the attenuation of the given IF or baseband filter as a function of the offset from the carrier. We assume that IFR already reaches its ultimate value at $f_{IF} + \Delta f$, and the attenuation of the interfering power reaching the detector due to Sp is roughly IFR.

NOTE: It is worth noting that, when measuring selectivity using Equation (2.9), we are not capable of quantifying the individual contributions of IFR, $SPRS$ and SBN. However, when designing a receiver, Equation (2.10) makes clear which circuits we should be focusing on.

The following computation proves Equation (2.10). From the discussion above, and with G the center-frequency gain, the total interference power $Ntot$ reaching the detector is

$$Ntot = \left[10^{-IFR/10} + 10^{SPRS/10} + 10^{SBN/10} \right] Sadj \cdot G \tag{2.17}$$

Referring again to Equation (2.1) and with the signal $Si = 2Sens$ applied to antenna input, just as in the CCR case, the receiver performance is as at sensitivity level, then $SNRd$ is roughly the same as at sensitivity. Thus

$$\frac{1}{SNRd} = \frac{Nd}{Sens \cdot G} \approx \frac{Ntot + Nd}{Si \cdot G} \tag{2.18}$$

Substituting $Si = 2Sens$ into (2.18) and using (2.5), (2.1) and (2.9) we get

$$10^{\frac{Sel}{10}} = \frac{Sadj}{Sens} = \frac{1}{SNRd \cdot [10^{-IFR/10} + 10^{SPRS/10} + 10^{SBN/10}]} \tag{2.19}$$

Taking 10 times the logarithm of Equation (2.19) completes the proof.

Example: Assume the following subsystem specifications:

- SSB noise at adjacent channel: $L = -112$ dBc/Hz.
- IF attenuation within adjacent channel: $IFR = 100$ dB.
- $CCR = -8$ dB.
- Channel bandwidth: $B = 150$ kHz.
- Synthesizer spurs: $SPRS \leq -85$ dB.

The selectivity computed with Equation (2.10) yields

$$Sel = -10 \log [10^{-10} + 10^{-8.5} + 10^{-11.2+5.2}] - 8 \approx 52 \ dB$$

If we neglect the SSB noise ($L = -\infty$), we get *Sel* $\approx$ 77 dB. This is a realistic example where the synthesizer phase noise is the limiting factor for the selectivity.

2.3.1.1 Computation of SHR Selectivity

We again refer to the SHR architecture defined in Figure 1.1 and Table 1.1. In this case, as in most narrowband receivers, the limiting factor for adjacent channel selectivity is the phase noise of the first LO. The reasons are:

- As explained in section 2.3.1, the LO, by the mixing action, causes any (even 'clean') interferer to acquire the very same LO 'noise tail'-shape characteristics. According to Equation (2.12), the LO noise 'tail' grows by about 6 dB/octave when approaching f_{LO}. The smaller the channel spacing, the closer the adjacent channel, and the closer the adjacent channel, the larger the noise tail power spread into the desired IF bandwidth. As noted previously, there is no protection against this interfering noise, since its frequency band falls exactly onto the operating channel.
- As pointed out before, for a given value of Δf, the LO noise increases when f_{LO} increases, thus the first LO is usually dominant as compared to the second LO, which has a much lower center frequency.
- In SHR receivers it is easy to obtain IF filters featuring extremely high rejection within adjacent channels. Moreover, there are reliable techniques to prevent occurrence of synthesizer spurs.
- Potential selectivity killers arising from parasitic interferences such as remodulation and microphonics, which we treat in Chapter 6, can also be kept low, especially in high-tier radios which are less cost-sensitive.

In view of the above, in most narrowband receivers, as far as selectivity is concerned, all the interferences, except LO phase noise, may be neglected, and Equation (2.17) can be approximated as

$$Sel \approx -SBN + CCR \ [dB] \qquad (2.20)$$

The adjacent channel selectivity for the SHR receiver of Figure 1.1, using the values given in Table 1.1, is readily computed using Equation (2.20) as shown in Figure 2.7. Specifically, with 25 kHz channel spacing, 18 kHz IF bandwidth and SSB noise level of -125 dBc/Hz at adjacent channel offset, we get

$$Sel = 125 - 10 \log_{10}(18 \cdot 10^3) - 10 \approx 72 \ dB \ @ \pm 25 \ kHz \qquad (2.21)$$

If the LO characteristics are unknown, but it is known that the SSB noise is the limiting factor in determining the selectivity, then we may deduce the simplified Leeson's equation (2.12) by just measuring the adjacent channel selectivity value as described in section 2.3.1.2.

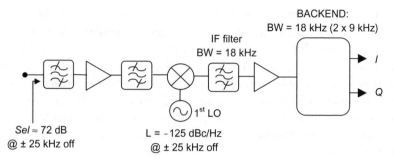

Figure 2.7 Computation of SHR narrowband adjacent channel selectivity

Indeed, once the value of *Sel* is known, we may compute *SBN* using Equation (2.20). Then the value of L at $\Delta f = \pm 25$ kHz is readily found from Equation (2.15), and thus the value of $A(f_{LO})$ in Equation (2.12) can be computed. At this point, Equation (2.12) together with the known values of B and *CCR*, allows the direct computation of selectivity for any arbitrary offset Δf.

2.3.1.2 Computation of DCR Selectivity

As we saw before, when using the IQ architecture, the band-pass RF bandwidth seen at the backend input is shared by the quadrature channels, each one of them being filtered by a low-pass filter half-bandwidth wide (see section 1.2.1.2).

Since in DCR there is no IF filter, the attenuation at adjacent channels depends entirely on the I and Q low-pass filters in the backend. Referring to a DCR receiver as in Figure 1.7 and Table 1.2, and by the considerations highlighted in section 2.7, let us compute the selectivity using the approximated Equation (2.20). With $CCR = -10$ dB, an SSB noise level of -135 dBc/Hz at 20 MHz offset and with 16 MHz signal pass-band we get (see Figure 2.8):

$$Sel = 135 - 10\log_{10}(16 \cdot 10^6) - 10 \approx 53 \; dB \, @ \pm 20 \; MHz \qquad (2.22)$$

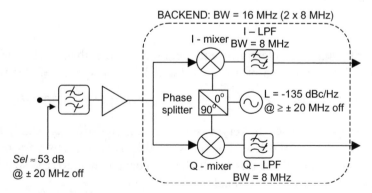

Figure 2.8 Computation of DCR broadband adjacent channel selectivity

Since the backend attenuation at the adjacent channel is 70 dB, we conclude that here too the LO phase noise is the limiting factor for selectivity. The main reason is that the on-chip oscillator cannot provide a phase noise level low enough, thus we are unable to take full advantage of the attenuation of the low-pass filters on-chip. However, if one is willing to use an external VCO instead of the one built on-chip, achieving an SSB noise level of -155 dBc/Hz @ 20 MHz offset is pretty straightforward. In this case, the backend I and Q low-pass filters will become the limiting factor, and neglecting oscillator spurs, we can expect a selectivity value computed with the full formula of Equation (2.10).

$$Sel = -10\,log\,[10^{-IFR/10} + 10^{SBN/10}] + CCR \; [dB] \qquad (2.23)$$

and specifically

$$Sel = -10\,log\,[10^{-70/10} + 10^{(-155+72)/10}] - 10 \approx 60 \; dB \qquad (2.24)$$

In this case the backend selectivity becomes the limiting factor, thus there is only a 7 dB improvement, so there is not much to be gained, unless the backend low-pass filters are made sharper.

2.3.1.3 Computation of VLIF Selectivity

The case of hardware-based VLIF is different from the previous two cases discussed. In view of the computation carried out in section 2.3.1.2, a glance at Table 1.3 is enough to conclude that the polyphase filter attenuation is the limiting factor for selectivity.

Then we ask ourselves: how 'bad' can we make the adjacent channel phase noise performance of the integrated VCO in this case? This is a proper question, since relaxation of SSB noise specs helps in reducing size, cost, complexity and current drain of the VCO (Chapter 5), which may be critical in low-tier applications. Again neglecting VCO spurs, the answer is easily provided by looking at Equation (2.23). If we assume that the selectivity figure is dominant by one order of magnitude, say

$$10^{-IFR/10} > 10 \times 10^{SBN/10} \Rightarrow SBN < -IFR - 10 \qquad (2.25)$$

we can verify that the selectivity will degrade by less than 1 dB due to VCO phase noise. Using the data of Tables 1.2 and 1.3 in inequality (2.25) we get

$$L < -10\,log_{10}(16 \cdot 10^6) - 45 = -117 \; dBc/Hz\,@|\Delta f| = 20 \; MHz \qquad (2.26)$$

Such a noise performance is very easy to obtain. Substituting (2.26) back into (2.23) with $CCR = -10$ dB, yields $Sel \approx 25$ dB. This figure is very poor, however, in many applications, this is all that is needed, while cost, size and consumption of the receiver are critical.

2.3.2 Measurement of Selectivity

The setup for the measurement of selectivity is identical to the setup for *CCR* described in Figure 2.6. The measuring procedure is identical to the one used for *CCR*, except:

- The selectivity should be measured at both sides of the desired channel, namely one channel lower and one channel higher in frequency. The interfering RF signal generator (B) should be tuned to either adjacent channels, and the test repeated for both. The worst-case result is taken as the selectivity spec. There is no guarantee that the selectivity will be the same or even similar on both sides, especially due to asymmetric synthesizer spurs.
- Unlike with *CCR*, selectivity measurements put severe constraints on the performance of the interfering signal generator B:
  - Generator B must have an SSB phase noise at least 10 dB better than the SSB phase noise of the synthesizer in the receiver under test. If not, we will measure a result worse that the actual Rx selectivity performance.
  - The spurious level of generator B at the adjacent channel must be at least 10 dB below the selectivity figure to be measured, otherwise we will again get a wrong result.
  - For generator A, as in the *CCR* case, there are no special requirements

Overlooking the performance of generator B is a common pitfall for RF engineers, and often yields puzzling measurement results. If an adequate generator cannot be found, the only thing we can say is that the Rx selectivity figure is not worse than the measured one.

2.4 Intermodulation Rejection

'Intermodulation' in the receiver context usually refers to third-order intermodulation rejection, which is the prime intermodulation interference that we should be concerned about. We refer the reader to Chapter 6 for a detailed analysis of intermodulation, and insight on intercept-point theory.

Taking again $Si = 2Sens$ and G the center-frequency gain as defined for *CCR*, consider two signals $R2\Delta$ and $R4\Delta$ both of the same RF power Sim, one of them centered at frequency $f_R + 2\Delta f$ (or $f_R - 2\Delta f$), and a second one centered at $f_R + 4\Delta f$ (or $f_R - 4\Delta f$), with f_R being the received frequency and Δf the channel spacing, as defined for selectivity measurement.

Set Sim at the smallest power value such that, when $R2\Delta$ and $R4\Delta$ are applied to the antenna port simultaneously with Si, the receiver output deteriorates to the performance obtained at sensitivity level. Then, the 'third-order intermodulation rejection', which we denote by $IMR3$, is the ratio, in dB, between the power values Sim and $Sens$, namely

$$IMR3 = 10\,log\,(Sim/Sens)\;[dB] \tag{2.27}$$

2.4.1 Computation of Intermodulation Rejection

The third-order intermodulation is due mainly to the fourth-order nonlinearity of the mixer(s) in the Rx front end (Chapter 6), and behaves as follows: increasing $R2\Delta$ and $R4\Delta$ by 1 dB, namely, increasing the value *Sim* by 1 dB, the interference generated at detector input increases by 3 dB. This is triggered by the presence, at the antenna port, of any couple of RF signals, $Rn\Delta$ and $R2n\Delta$, one of them at frequency $f_R + n\Delta f$ (or $f_R - n\Delta f$) and the other at $f_R + 2n\Delta f$ (or $f_R - 2n\Delta f$), where n is an integer. This is because the fourth-order nonlinear mixer(s) term

$$\{A\cos[2\pi(f_R \pm n\Delta f)t] + B\cos[2\pi(f_R \pm 2n\Delta f)t] + C\cos(2\pi f_{LO}t)\}^4$$

where f_{LO} is the local oscillator frequency, produces, among others, a signal proportional to $A^2BC \times \cos(2\pi f_{IF}t)$, which is a product at IF frequency f_{IF}, which is indistinguishable from the wanted signal after mixing. If A and B are signals of equal amplitude, and since C is fixed, the spurious signal is proportional to A^3. The result is a noise signal power $N3$ that adds up to the noise at detector input.

The critical figure in determining the intermodulation is a receiver parameter called 'third-order input intercept point', in short, $IP3i$ [dBm].

Once $IP3i$ is known, then $IMR3$ is computed using (Chapter 6):

$$IMR3 = \tfrac{2}{3}(IP3i - Sens|_{dBm}) + \tfrac{1}{3}CCR \qquad (2.28)$$

Conversely, once *Sim* has been measured according to the procedure described in the previous section, then $IP3i$ can be readily computed substituting (2.27) into (2.28). Note, however, that if the radio front end provides some attenuation at the frequencies of $Rn\Delta$ and $R2n\Delta$, then $IP3i$ will be dependent on that attenuation, thus $IP3i$ may become frequency dependent.

The following computation proves Equation (2.28). With $Si = 2Sens$ present, and *Sim* adjusted at a level such that the receiver performs as at sensitivity, as in the CCR case, the noise $N3$ must be roughly equal to Nd, thus we get, with the help of Equation (2.1)

$$\frac{N3 + Nd}{2Sens \cdot G} \approx \frac{1}{SNRd} \rightarrow \frac{N3}{Sens \cdot G} \approx \frac{1}{SNRd} \qquad (2.29)$$

Using logarithmic notation and with the help of Equation (2.5), (2.29) roughly yields

$$N3|_{dBm} = G|_{dB} + Sens|_{dBm} + CCR \qquad (2.30)$$

The value of $IP3i$ is defined as follows: if we apply, at antenna input, a couple of interfering signals $Rn\Delta$ and $R2n\Delta$ of power $IP3i$ each, they will generate, at detector input, the same power $IP3o$, in dBm, as if we applied, at antenna input, a single on-channel signal of power $IP3i$.

With G being the center-frequency gain from antenna to detector input, it follows from the definition of $IP3i$ that

$$IP3o = IP3i + G|_{dB} \qquad (2.31)$$

We show in Chapter 6 that a couple of interfering signals $Rn\Delta$ and $R2n\Delta$ of power Sim, generates at the detector input an interference $N3$ of power

$$N3|_{dBm} = IP3o - 3(IP3i - Sim|_{dBm}) \qquad (2.32)$$

Substituting (2.31) in (2.32) and equating it to (2.30), yields

$$IP3i - 3(IP3i - Sim|_{dBm}) = 3Sens|_{dBm} - 2Sens|_{dBm} + CCR \qquad (2.33)$$

finally getting

$$(Sim|_{dBm} - Sens|_{dBm}) = IMR3 = \tfrac{2}{3}(IP3i - Sens|_{dBm}) + \tfrac{1}{3}CCR$$

Example: Assume the following specifications for a GSM cellphone:

- $Sens = -104.5$ dBm
- $CCR = -8$ dB
- $IP3i = -10$ dBm

The third-order intermodulation computed with (2.28) yields

$$IMR3 = \tfrac{2}{3}(-10 + 104.5) - \tfrac{8}{3} \approx 60.3 \; dB$$

Notes: $IP3i = -10$ dBm is the typical natural performance achieved using bipolar active mixers. Low-noise amplifiers (LNA) and other RF circuits may also contribute to intermodulation, each one with a different $IP3i$ value. If no one of them has dominant contribution, then the overall receiver $IP3i$ will be some mathematical combination of the individual $IP3i$ values.

2.4.2 Measurement of Intermodulation Rejection

The setup for the measurement of third-order intermodulation rejection is described in Figure 2.9. Here we show only the setup with digital modulation. The measurement using analog modulation is readily inferred from Figure 2.5.

First the interfering generators (B and C) are shut off, and the level of the main generator (A) is increased until the output from the receiver fits the sensitivity performance (note that due to the attenuation of the combining network, the main signal level will be higher than sensitivity level). Then, the level of the main generator is increased by 3 dB, causing the output performance (BER or SINAD) to be better than the value at sensitivity. Next, the interfering RF signal generator (C) is FM modulated either with an analog or a digital signal, while the interfering RF signal generator (B) is left unmodulated.

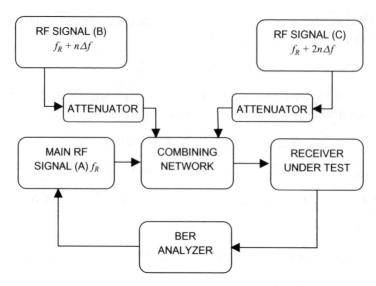

Figure 2.9 *IMR*3 measurement with digital modulation

The reason why only the far-out signal should be modulated is that *IMR*3 involves the square power of the nearer signal, which would cause its FM modulation index to double, thus exceeding the channel bandwidth.

Now, the level of the interfering generators (B and C) is increased until the receiver output performance deteriorates to the sensitivity value.

The *IMR*3 is the ratio, in dB, between the (equal) power of generators B and C and the power level of generator A at this stage. Care should be taken in the following:

- The modulation of signal C should be chosen so that it cannot be confused with the modulation of signal A. For instance, in analog measurements, if signal A is modulated with a 1000 Hz sinusoid, then signal C will be modulated with a 400 Hz sinusoid.
- The modulation level of signal C should be such that it does not cause the RF signal to exceed the channel bandwidth, otherwise part of the interfering power will be lost, and the measurement will be incorrect.

While there are no special requirements on generator A, intermodulation rejection is a very demanding test as far as signal generators B and C are concerned.

Several points are worth mentioning:

- Some measurement methods are defined with $n = 1$ (namely, the interfering signals are at one channel and two channels off). A quick glance at the selectivity measurement procedure reveals that with $n = 1$, the *IMR*3 result can never be better than the selectivity figure whenever the limiting factor is the synthesizer phase noise, or the phase noise of signal generator B (the one closer to the channel frequency f_R). Thus, $IMR3$ measurements performed using $n = 1$ are somewhat meaningless. This is because usually the $IMR3$ figure is much better than the selectivity, so that we end up measuring again the selectivity value. This is why it became customary to use $n = 2$, as described at the beginning of this section, but even then, the relief is not major, since the synthesizer (or generator) phase noise effect is reduced by only 6 dB when n doubles (Chapter 5). The best approach of measuring the true $IMR3$ value is to adopt the largest value of n, for which the signal at frequency $f_R + 2n\Delta f$ is still within the bandwidth of the RF front filters, and suffers no attenuation. The more far out the frequency $f_R + n\Delta f$, the less the phase noise of the signal generator and of the Rx synthesizer will affect the $IMR3$ measurement.

- Since the $IMR3$ figure is usually better than selectivity, even with relatively large values of n, still care must be taken in using low-noise signal generators for the interfering signals.
- The attenuators (≥ 10 dB attenuation each) in the setup of Figure 2.9 are an absolute requirement. This is because generators B and C are usually built so that, when delivering relatively high-power levels they work without internal attenuators, by controlling the RF level to the driver of the output stage. Thus, unless external attenuators are used, we will end up measuring the $IMR3$ of the signal generators, which is usually much worse than the $IMR3$ of the receiver. This is a very common mistake that leads to wrong and puzzling results. In contrast, the on-channel generator (A) works at very low RF levels, near sensitivity, thus it is normally well protected by its internal attenuator.

2.5 Half-IF Rejection

This specification is not relevant for direct conversion receivers (DCR) whose IF frequency is zero (for DCR receivers, as we see later, there is an equivalent spec called 'second-order distortion rejection').

Take again $Si = 2Sens$ and G the center-frequency gain as defined for CCR, denote the center frequency by f_R, and the IF frequency (including the VLIF case) by f_{IF}. Consider a signal of power $S_{IF/2}$ at frequency $f_{IF/2}$

$$f_{IF/2} = f_R \pm \tfrac{1}{2} f_{IF}, \quad f_{LO} = f_R \pm f_{IF} \qquad (2.34)$$

where f_{LO} denotes the local oscillator frequency, and the sign $\pm$ is for the upper side of lower-side LO injection.

Set $S_{IF/2}$ at the smallest power value such that, when applied to the antenna port simultaneously with Si, the receiver output deteriorates to the performance obtained at sensitivity level. Then, the Rx 'half-IF rejection', which we denote by $HIFR$, is the ratio, in dB, between the power values $S_{IF/2}$ and $Sens$, namely

$$HIFR = 10 \log(S_{IF/2}/Sens) \ [dB] \qquad (2.35)$$

2.5.1 Computation of Half-IF Rejection

The half-IF rejection (*HIFR*) behaves as follows: increasing $S_{IF/2}$ by 1 dB, the interference generated at detector input increases by 2 dB.

The half-IF spur is due mainly to the fourth-order nonlinearity of the mixer(s) in the Rx front end. Nevertheless, we refer to it as a second-order intermodulation phenomenon, since the disturbance grows (in dB) twice as fast as the interfering signal. The mechanism generating half-IF spurs is as follows: when the LO and half-IF frequency as described in Equation (2.35) reach the mixer(s), the fourth-order nonlinear term

$$\{A \cos[2\pi(f_R \pm f_{IF/2})t] + B \cos[2\pi(f_R \pm f_{IF})t]\}^4$$

produces, among the others, a spurious signal proportional to $A^2 B^2 \times \cos(2\pi f_{IF}t)$, which is at the IF frequency, and thus indistinguishable from the wanted signal after mixing. Since the LO amplitude B is fixed, the spurious signal is proportional to A^2.

The critical figure in determining the half-IF rejection is a receiver parameter called 'second-order input intercept point', in short, *IP2i* [dBm], and the result is a noise signal, of power $N2$, that adds up to the noise at detector input. Once *IP2i* is known, then *HIFR* is computed using

$$HIFR = \tfrac{1}{2}(IP2i - Sens|_{dBm}) + \tfrac{1}{2}CCR \qquad (2.36)$$

Conversely, once $S_{IF/2}$ has been measured according to the procedure described in the previous section, then *IP2i* can be readily computed substituting (2.35) into (2.36). Note, however, that if the radio front end provides some attenuation at the frequency $f_{IF/2}$, then *IP2i* will be dependent on that attenuation, thus *IP2i* may become frequency dependent.

The following computation proves Equation (2.36). With $Si = 2Sens$ present, and $S_{IF/2}$ adjusted at a level such that the receiver performs as at sensitivity, as in the *CCR* case, the noise power $N2$ must be roughly equal to Nd, thus we get, with the help of Equation (2.1)

$$\frac{N2 + Nd}{2Sens \cdot G} \approx \frac{1}{SNRd} \rightarrow \frac{N2}{Sens \cdot G} \approx \frac{1}{SNRd} \qquad (2.37)$$

Using logarithmic notation and with the help of Equation (2.5), (2.37) roughly yields

$$N2|_{dBm} = G|_{dB} + Sens|_{dBm} + CCR \qquad (2.38)$$

The value of *IP2i* is defined as follows: if we apply, at antenna input, an interfering signal $S_{IF/2}$ of power *IP2i*, it will generate, at detector input, the same power *IP2o*, in dBm, as if we applied, at antenna input, an on-channel signal of power $IP2i$. With G being the center-frequency gain from antenna to detector input, it follows from the definition of $IP2i$ that:

$$IP2o = IP2i + G|_{dB} \qquad (2.39)$$

By the definition of the second-order intermodulation phenomena, it follows that an interfering signal of power $S_{IF/2}$, generates at the detector input an interference $N2$ of power

$$N2|_{dBm} = IP2o - 2(IP2i - S_{IF/2}|_{dBm}) \qquad (2.40)$$

Substituting (2.39) in (2.40) and equating it to (2.38), yields

$$IP2i - 2(IP2i - S_{IF/2}|_{dBm}) = 2Sens|_{dBm} - Sens|_{dBm} + CCR \qquad (2.41)$$

finally getting

$$(S_{IF/2}|_{dBm} - Sens|_{dBm}) = HIFR = \tfrac{1}{2}(IP2i - Sens|_{dBm}) + \tfrac{1}{2}CCR$$

NOTE: Other RF circuits may also contribute to half-IF spurs, each one with a different $IP2i$ value. If no one of them has a dominant contribution, then the overall receiver $IP2i$ will be some mathematical combination of the individual $IP2i$ values.

Example: Assume we measured the following for a GSM cellphone:

- $Sens = -104.5$ dBm
- $CCR = -8$ dB
- $HIFR = 60$ dB

Then, using Equation (2.36) we compute the second-order input intercept point

$$IP2i = 2HIFR - CCR + Sens|_{dBm} = 120 + 8 - 104.5 = +23.5 \ dBm$$

Notes: The $HIFR$ figure of 60 dB is usually obtained with the help of a front filter attenuating signals at frequency $f_{IF/2}$, which can be used only with super-heterodyne. This filter is required since the natural $IP2i$ achieved using bipolar mixers is around -7 dBm, leading to about 45 dB $HIFR$. The latter value is the performance level in VLIF architectures where no front filter can be added. Again, VLIF architectures are intrinsically low-performance.

2.5.2 Measurement of Half-IF Rejection

The setup for the measurement of half-IF rejection is identical to the setup for CCR described in Figure 2.6. The measuring procedure is identical to the one used for CCR, except:

- The interfering RF signal generator (B) is tuned to

$$f_{IF/2} = f_R \pm \tfrac{1}{2} f_{IF}$$

  with the $\pm$ sign set according to whether the radio uses upper- or lower-side injection.
- In the case of heterodyne, there are no special constraints on the interfering RF signal generator (B).
- In the VLIF case, since the interfering signal is close to the channel frequency, all the restrictions on SSB noise and spurious required for the measurement of selectivity (section 2.3.2) apply to RF signal generator B.

2.6 Image Rejection

This specification is not relevant for direct conversion receivers (DCR) whose IF frequency is zero. The image response of a receiver is a phenomenon inherent to mixer(s) functionality.

Take again $Si = 2Sens$ and G the center-frequency gain as defined for CCR. Denote the center frequency by f_R, and the IF frequency (including the VLIF case) by f_{IF}. Consider a signal of power S_{Image} at frequency f_{Image}

$$f_{Image} = f_R \pm 2 f_{IF} \qquad (2.42)$$

with the $\pm$ sign set according to whether the radio uses upper- or lower-side injection.

Set S_{Image} at the smallest power value such that, when applied to the antenna port simultaneously with Si, the receiver output deteriorates to the performance obtained at sensitivity level. Then, the 'image rejection', which we denote by IR, is the ratio, in dB, between the power values S_{Image} and $Sens$, namely

$$IR = 10 \, log \, (S_{Image}/Sens)[dB] \qquad (2.43)$$

The image spur is mainly due to the second-order nonlinearity of the mixer(s) in the Rx front end. Nevertheless, we refer to it as a first-order phenomenon, since the disturbance grows exactly as fast as the interfering signal. The mechanism generating the image spur is the same that generates the desired mixing action: when the LO and image frequency reach the mixer(s), the second-order nonlinear term

$$\{A \, cos [2\pi (f_R \pm f_{Image})t] + B \, cos [2\pi (f_R \pm f_{IF})t]\}^2$$

produces, among others, a spurious signal proportional to $AB \times cos(2\pi f_{IF}t)$, which is at the IF frequency, and thus indistinguishable from the wanted signal after mixing.

Since the LO amplitude B is fixed, the spurious signal is proportional to A. Being of first order, the image spur bears no natural rejection at all. If we measure it without additional protection at front-end level, we come up with the same value as the *CCR*, namely, around -10 dB.

NOTE: *IR* figures above 30 dB are usually obtained with the help of a front filter attenuating signals at $2f_{IF}$ offset, which can be used only with superheterodyne. In VLIF radios, where front filtering cannot provide image rejection, special architectures are used, which are able to provide an *IR* of about 35 dB. Again we see that VLIF receivers are intrinsically low performance.

2.6.1 Computation of Image Rejection

As far as the mixer(s) is concerned, a signal at frequency f_{Image} is undistinguishable from an on-channel signal, and it results in a noise signal of power N_{Image} that adds up to the noise at detector input. The only protection against it is the attenuation of a front-end filter or the rejection of an image-reject mixer, both of which are denoted by A_{Image}. The image rejection *IR* is computed using

$$IR = 10 \log (A_{Image}) + CCR \ [dB] \tag{2.44}$$

The following computation proves Equation (2.44): again with G the center-frequency gain from antenna to detector, and with S_{Inage} at the antenna port, the noise power reaching the detector input is

$$N_{Image} = G \cdot S_{Image} / A_{Image} \tag{2.45}$$

As in the *CCR* case, the noise N_{Image} must be roughly equal to Nd, thus we get, with the help of Equation (2.1)

$$\frac{N_{Image} + Nd}{2Sens \cdot G} \approx \frac{1}{SNRd} \rightarrow \frac{N_{Image}}{Sens \cdot G} \approx \frac{1}{SNRd} \tag{2.46}$$

Substituting (2.45) in (2.46) yields

$$\frac{S_{Image}}{Sens} \approx \frac{A_{Image}}{SNRd} \tag{2.47}$$

Computing (2.47) in dB, using (2.5) and substituting (2.43), leads to (2.44).

2.6.2 Measurement of Image Rejection

The setup for the measurement of image rejection is identical to the setup for *CCR* described in Figure 2.6. The measuring procedure is identical to the one used for *CCR*, except:

- The interfering RF signal generator (B) is tuned to:

$$f_{Image} = f_R \pm 2f_{IF}$$

with the $\pm$ sign set according to whether the radio uses upper or lower-side injection.

- In the case of heterodyne, there are no special constraints on the interfering RF signal generator (B).
- In the VLIF case, since the interfering signal is close to the channel frequency, all the restrictions on SSB noise and spurious content required for the measurement of selectivity (section 2.3.2) apply to RF signal generator B.

2.7 Second-Order Distortion Rejection

This specification is relevant for direct conversion receivers (DCR) whose IF frequency is zero. The phenomenon is triggered by the presence of a strong interferer anywhere near the center frequency, and produces baseband noise at the demodulator input. Since the IF of a DCR is at zero frequency, this is analogous to the half-IF spur for heterodyne and VLIF receivers.

Take again $Si = 2Sens$ and G the center-frequency gain as defined for CCR. Denote the center frequency by f_R, and consider a signal of average power S_{D2} at any frequency f_{D2}

$$f_{D2}(t) = f_R + \Delta f(t) \tag{2.48}$$

where $\Delta f(t)$ is any variable low-frequency offset fluctuation for which the signal is not attenuated by the RF chain filters.

Set S_{D2} at the smallest power value such that, when applied to the antenna port simultaneously with Si, the receiver output deteriorates to the performance obtained at sensitivity level. Then, the 'second-order distortion rejection', which we denote by $D2R$, is the ratio, in dB, between the power values S_{D2} and $Sens$, namely

$$D2R = 10\log(S_{D2}/Sens) \ [dB] \tag{2.49}$$

2.7.1 Computation of Second-Order Distortion Rejection

The second-order distortion rejection behaves as follows: increasing S_{D2} by 1 dB, the interference generated at detector input increases by 2 dB.

The second-order distortion rejection of a receiver is caused mainly by the second-order nonlinearity of the RF chain, namely LNA, mixer(s) and, in the case of dual-conversion receivers, also amplifiers in the first IF chain. With a strong interferer $S_{D2} = A(t)\cos(2\pi f_{D2}(t) \cdot t)$ present at antenna input, a second-order nonlinear term yields

$$\{A(t)\cos(2\pi f_{D2}(t) \cdot t)]\}^2 = \tfrac{1}{2}[A(t)]^2 + \tfrac{1}{2}[A(t)]^2 \cos(4\pi f_{D2}(t) \cdot t)$$

where $A(t)$ is a low-frequency signal. Note that, as opposed to the half-IF and third-order intermodulation cases, the LO is not involved with the generation of this spur. This is apparent from the fact that f_{D2} may have an arbitrary value within the front-end bandwidth.

Since $A(t)$ is low frequency, we get a spurious signal proportional to the amplitude $A^2(t)$, which is at baseband frequency, and thus indistinguishable from the wanted signal after direct conversion mixing. Note that the spur bandwidth corresponds to the convolution of the Fourier transform of $A(t)$ with itself. This yields a spectrum twice as wide as the bandwidth of $A(t)$, and therefore the spectra of the spur may exceed the bandwidth of the baseband filter, making the definition of the $D2R$ dependent on the bandwidth of $A(t)$, not only on its power. Note also that the bandwidth of $A(t)$ should not be too small since, in general, DCR architectures notch-out the 'DC' frequency in the baseband, which may again result in an ambiguous definition of $D2R$. To avoid ambiguity, the bandwidth of $A(t)$ should be half the baseband bandwidth.

The critical figure, as in the half-IF rejection, is the receiver parameter called 'second-order input intercept point' or, in short, $IP2i$ [dBm], and the result is a noise signal, of power $N2$, that adds up to the noise at detector input.

Once $IP2i$ is known, then $D2R$ is computed as in the half-IF case using

$$D2R = \tfrac{1}{2}(IP2i - Sens|_{dBm}) + \tfrac{1}{2}CCR \qquad (2.50)$$

The proof of (2.50) is identical to the proof of (2.36) replacing

$$S_{IF/2}|_{dBm} \Rightarrow S_{D2}|_{dBm} \text{ and } HIFR \Rightarrow D2R$$

NOTE: There is no front-end protection for this spur, therefore, in a DCR the linearity requirements on the active RF components are much more severe than for heterodyne receivers, which reflects both in price and current consumption.

2.7.2 Measurement of Second-Order Distortion Rejection

The setup for the measurement of second-order distortion rejection is identical to the setup for CCR described in Figure 2.6. The measuring procedure is identical to the one used for CCR, except for:

- The interfering RF signal generator (B) is tuned somewhere within the front filter pass-band, and as far as possible from the center frequency, so as to avoid LO noise effects already seen in selectivity measurement (see section 2.3). With this choice, there are no special constraints on the interfering RF signal generator (B).
- The interfering RF signal generator (B) should be AM-modulated with a modulating signal $A(t)$ of bandwidth half of the receiver baseband bandwidth.

NOTE: Since the $D2R$ is a function of the interferer signal amplitude only, AM-modulation is all we need for measurement.

2.8 Blocking

The Rx 'blocking' phenomenon is triggered by the presence of one or more strong interferers anywhere within the front-end pass-band, and away from channel frequency. Blocking is one of the worst 'system killers' in cellular and other densely populated RF environments, and can be caused by unknown and alien nearby systems. Often it cannot be predicted or planned ahead of time, and may lead to a total system paralysis. Apart from a spurious-free and low-noise-floor synthesizer, there is no protection against it.

Take again $Si = 2Sens$ and G the center-frequency gain as defined for CCR. Denote the center frequency by f_R, and consider a signal of power S_{Block} at any frequency f_{Block} such that

$$f_{Block} = f_R + n\Delta f \qquad (2.51)$$

where Δf is the channel width, and n is a large integer for which f_{Block} is still within the front-end pass-band (f_{Block} should be several channels away from f_R, but not attenuated by the front-end filters).

Set S_{Block} at the smallest power value such that, when applied to the antenna port simultaneously with Si, the receiver output deteriorates to the performance obtained at sensitivity level. Then, the Rx 'blocking', which we denote by $Block$, is the ratio, in dB, between the power values S_{Block} and $Sens$, namely

$$Block = 10\,log\,(S_{Block}/Sens)\ [dB] \qquad (2.52)$$

2.8.1 Computation of Blocking

The blocking is due to receiver LO far-out SSB noise floor $L(\infty)$ and synthesizer spurs (see section 2.3) interacting with an interferer away from channel frequency and producing a noise signal N_{Block} at demodulator input.

Just as in the case of selectivity, even if the interferer consists of a perfectly clean 'delta' function, the synthesizer phase noise and spurs modulate it within the mixer, causing its spectrum to acquire the very same spectral shape of the LO, and generating wide 'tails' that spread out towards the center frequency.

If there are several strong interferers away from the channel frequency, as is often the case in a crowded RF environment, their 'tails' sum up together and produce noise at demodulator input. For a single interferer, the blocking is

$$Block = -10\,log\,[10^{SPRS/10} + 10^{SBN/10}] + CCR\ [dB] \qquad (2.53)$$

where SBN is defined in (2.15) and $SPRS$ in (2.16).

The proof of (2.53) is identical to the one for selectivity in section 2.3.1, setting $IFR = \infty$ and replacing in equations (2.10) through (2.19) and in the associated paragraphs:

$$Sel \Rightarrow Block,\ f_0 + \Delta f \Rightarrow f_{Block}$$

$$Sadj \Rightarrow S_{Block},\ Ntot \Rightarrow N_{Block} \qquad (2.54)$$

Example: It is of interest to compare the single-interferer blocking performance of two implementations of a GSM phone: the first using a standalone VCO with a noise floor $L = -155$ dBc/Hz, and the other using the system-on-chip VCO with a noise floor of $L = -130$ dBc/Hz.

Recalling that in GSM the IF bandwidth is $B = 140$ kHz and $CCR = -8$ dB, and assuming that spurs are negligible, we get, using (2.53)

$$Block = -10 \log (10^{SBN/10}) + CCR = -[L + 10 \log (B)] + CCR$$

which yields

$$Block|_{standalone} = 95.5 \; dB$$

$$Block|_{on-chip} = 70.5 \; dB$$

It is apparent that, basing the design on a system-on-chip, the built-in on-chip VCO cannot be used for high-performance radios.

2.8.2 Measurement of Blocking

The setup for the measurement of blocking is identical to the setup for *CCR* described in Figure 2.6. The measuring procedure is identical to the one used for *CCR*, except:

- The blocking should be measured at a frequency f_{Block} several channels away from f_R, but not attenuated by the front-end filters (usually $n > 10$).
- If the far-out spectral picture of the LO is unknown, then the interfering signal *B* generating S_{Block} must be swept all over the front-end pass-band to catch the effects of unknown synthesizer spurs and noise 'bumps' (Chapter 6).
- Generator B must have far-out SSB phase noise and spurs at least 10 dB better than the SSB phase noise of the synthesizer in the receiver under test, otherwise, we will measure a result worse that the actual blocking. However, since the interferer is far from f_R, this can be easily achieved by coupling the output of generator B through a band-pass filter.

2.9 Dynamic Range

Receiver RF circuits cannot operate with indefinitely large input signals at the antenna port. When a desired on-channel signal becomes too large, the receiver performance will begin to deteriorate, and at a certain input power level, it will worsen back to the sensitivity performance.

Denote by *Sat* the largest RF power at the antenna input and at center frequency that, when applied to the antenna without other interfering RF signals, will cause the receiver output to deteriorate to the performance obtained at sensitivity level.

The Rx 'dynamic range' *DR* is the ratio, in dB, between *Sat* and the sensitivity power level, namely

$$DR = 10 \log (Sat/Sens) \; [dB] \tag{2.55}$$

The dynamic range is mainly the result of a strong RF signal driving one or more circuits out of the proper operating/bias point, either by generating parasitic

DC currents due to rectification of the RF power, which in turn change circuit bias, or by driving one or more output stages into saturation.

2.9.1 Computation of Dynamic Range

There is no closed-form formula that fits every case, and the dynamic range is very dependent on receiver architecture and modulation type. For instance, in FM receivers, where deep circuit saturation is acceptable, there is virtually no limit to dynamic range. As opposed to that, in dense-constellation QAM modulation schemes, requiring highly linear receiver line-up, the dynamic range may be as low as 60 dB, and adaptive attenuators at antenna port may be required for proper operation.

Often the stages located closer to the detector are the more critical, since the gain from antenna port increases through the line-up towards the detector, and thus the signal level becomes higher making saturation more likely to occur.

2.9.2 Measurement of Dynamic Range

The measurement setup for dynamic range is identical to the one used for sensitivity (section 2.1.2). The procedure is also the same, except that the measurement is done twice:

- First measure the sensitivity *Sens*.
- Then increase the input signal until the receiver output performance deteriorates to the sensitivity value. The signal generator power at this stage is the value of *Sat*.
- Then compute *DR* according to Equation (2.55).

2.10 Duplex Desense

In true-duplex radios (as opposed to TDD – time division duplex), receiver and transmitter operate simultaneously at different frequencies, and the receiver input is isolated from the transmitter power by either a duplexer or by virtue of two separated Tx and Rx antennas. However, the Tx carrier 'tail' at Rx frequency, due to synthesizer SSB phase noise (section 2.3), may physically reach the receiver input causing deterioration of the signal-to-noise ratio *SNRd* at the detector. To restore the *SNRd* value required for sensitivity performance, the power of the desired signal must be increased, which results in an effective receiver sensitivity degradation.

Again let *Sens* be the receiver sensitivity measured with the transmitter disabled, and denote by *TSens* the receiver sensitivity measured with the transmitter operating simultaneously, which we refer to as 'duplex sensitivity'. We define the duplex desense *DS* as

$$DS = 10 \, log \, (TSens/Sens) \; [dB] \qquad (2.56)$$

2.10.1 Computation of Duplex Desense

Since the Tx-Rx physical isolation is not perfect, some power from the transmitter tail, stretching out to Rx center frequency, reaches the Rx input resulting in a noise power $N_{Desense}$ which adds up at detector input. The mechanism is either the duplexer leakage, when one common antenna is used, or the direct electromagnetic coupling between Tx and Rx antennas, when separate antennas are used. In the analysis that follows we refer to section 2.3.1. Moreover we assume that the Tx carrier power is attenuated enough so it does not cause receiver blocking (section 2.8).

Let:

- G: the power gain from Rx input to detector input.
- A_{TR}: Tx port output to Rx port input attenuation at Rx center frequency.
- S_T [watt]: the transmitter power at Tx frequency.
- B [Hz]: the receiver bandwidth.
- L [dBc/Hz]: the SSB phase noise of the Tx synthesizer at Rx frequency.

The phase noise PHN_T generated by the transmitter at Rx center frequency is computed using Equation (2.13), substituting S_T for S_{LO}

$$PHN_T = 10^{\frac{L}{10}} S_T \ [watt/Hz] \tag{2.57}$$

With the isolation A_{TR}, the noise $N_{Desense}$ reaching the detector input is

$$N_{Desense} = PHN_T \cdot B \cdot G/A_{TR} = 10^{\frac{L}{10}} S_T \cdot B \cdot G/A_{TR} \tag{2.58}$$

It follows from (2.1) that, for sensitivity-level performance we need

$$\frac{1}{SNRd} = \frac{Nd}{Sens \cdot G} = \frac{Nd + N_{Desense}}{TSens \cdot G} \tag{2.59}$$

Again using (2.1) to substitute $Nd = Sens \cdot G/SNRd$ in the above, we may compute (2.56) directly from (2.59), however the computation is cumbersome. In order to 'feel' the order of magnitude of the problem, let us compute the required isolation if we allow a (mild) 3 dB sensitivity degradation. For this purpose we assume

$$N_{Desense} = Nd \tag{2.60}$$

which, substituting in (2.59) yields $DS = 3$ dB. Then, substituting (2.1) and (2.58) in (2.60), we get

$$Sens/SNRd = 10^{\frac{L}{10}} S_T \cdot B/A_{TR} \tag{2.61}$$

Using the sideband noise value SBN as defined in (2.15), and the CCR as defined in (2.5) Equation (2.61) finally yields

$$A_{TR}|_{dB} = SBN + \frac{S_T}{Sens}\bigg|_{dB} - CCR \tag{2.62}$$

Example: In a typical scenario for:

- $S_T = +27$ dBm
- $Sens = -115$ dBm
- $B = 18$ kHz
- $CCR = -8$ dB
- $L = -150$ dBc/Hz at Rx frequency

Equation (2.62) gives a required Tx port to Rx port attenuation of

$$A_{TR} = -150 + 10 \log (18 \cdot 10^3) + 27 + 115 + 8 \approx 42.5 \; dB$$

The above attenuation is readily attained using a duplexer, but is very difficult to achieve with the use of two separate antennas. As pointed out before, in a portable device, antenna-to-antenna isolation is of the order of 15 dB. To achieve greater isolation, vertical pole-top physical separation is required.

2.10.2 Measurement of Duplex Desense

The measurement setup for digital modulation and using duplexer separation, is according to Figure 2.10. The attenuator is required in order to prevent the transmitter power from damaging the RF signal generator. However, since we are interested in the ratio *TSens/Sens*, there is no need to subtract the attenuator value from the level reading of the signal generator. The procedure is according to the following:

- Switch OFF the transmitter and record the generator level $S1$ (in dBm) that yields sensitivity performance using the sensitivity procedure described in section 2.1.2 for measuring *Sens*.
- Switch ON the transmitter and record the generator level $S2$ (in dBm) that yields sensitivity performance using the sensitivity procedure described in section 2.1.2 for measuring *Sens*. Here $S2$ is the duplex-sensitivity *TSens*.
- Compute the desense:

$$DS|_{dB} = S2|_{dBm} - S1|_{dBm} \qquad (2.63)$$

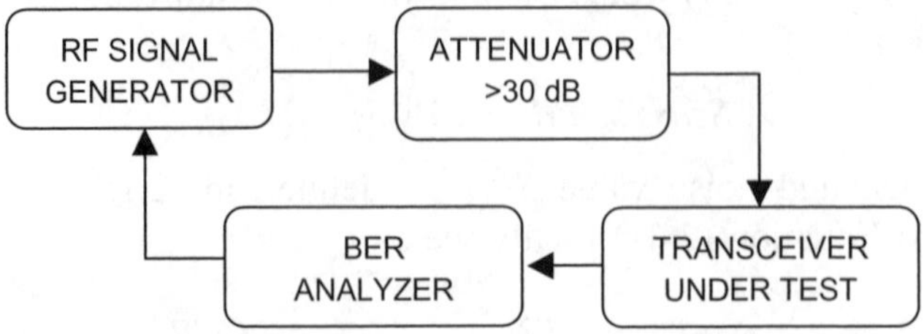

Figure 2.10   Duplex desense measurement with duplexer

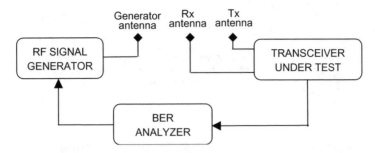

Figure 2.11 Duplex desense measurement with separate antennas

When using two separate antennas, the test must be done 'over the air' with the setup of Figure 2.11.

2.11 Duplex-Image Rejection

The duplex image is an Rx interference generated in true-duplex radios.

Denote the receiver center frequency by f_R, and the transmitter center frequency by f_T. The duplex-image frequency f_{DI} is the mirror of f_R with respect to f_T, namely

$$f_{DI} = f_T + (f_T - f_R) = 2f_T - f_R \qquad (2.64)$$

Let *TSens* be the receiver duplex sensitivity as defined in section 2.10. Take $TSi = 2TSens$, and consider a signal of power S_{DI} at frequency f_{DI}.

Set S_{DI} at the smallest power value such that, when applied to the antenna port simultaneously with *TSi*, and while transmitting rated RF power at frequency f_T, the receiver output deteriorates to the performance obtained at duplex sensitivity level. Then, the 'duplex-image rejection', which we denote by *DIR*, is the ratio, in dB, between the power values S_{DI} and *TSens*, namely

$$DIR = 10\,log\,(S_{DI}/TSens)\ [dB] \qquad (2.65)$$

2.11.1 Computation of Duplex-Image Rejection

The duplex-image interference can be generated by several mechanisms. The most obvious, which we do not analyze here, is when, due to insufficient Tx–Rx isolation, the transmitter signal reaches the receiver input with insufficient attenuation, and acts as a parasitic LO signal. Under this condition, the LNA or the first mixer may act as an harmonic mixer.

One less evident mechanism is when a signal at frequency f_{DI} reaches the output of a Tx power amplifier (PA). Although the exact analysis of the phenomena is complex and strongly dependent on the characteristics of the PA, we can reach a qualitative understanding and a feeling of the order of magnitude of the interference by using the simple model described in Figure 2.12 for duplexer

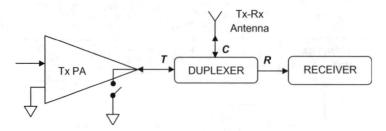

Figure 2.12   Duplex-image rejection with duplexer operation

operation. A very similar analysis holds for the case of separate antennas. For the sake of simplicity, we assume that the output impedance of the PA exhibits the behavior of a switch closing periodically at transmit frequency rate, either shorting to ground the duplexer terminal T, or leaving it disconnected. This is not a bad guess, especially with FM transmitters. From the above it follows that if a signal at frequency $\omega_{DI} = 2\pi f_{DI}$, coming from the common antenna terminal C reaches the PA output with voltage V_{DI}, its instantaneous value

$$v_{DI}(t) = V_{DI} \cos(\omega_{DI} t)$$

will be multiplied by the 'chopping' function:

$$S_\tau(t) = \sum_{n=-\infty}^{\infty} [u(t - nT) - u(t - \tau - nT)], \, T = \frac{1}{f_T}, 0 \leqslant \frac{\tau}{T} \leq 1 \qquad (2.66)$$

with the Heaviside function given by

$$u(t) = \begin{cases} 0, & t \leq 0 \\ 1, & t > 0 \end{cases} \qquad (2.67)$$

and $S_\tau(t)$ is shown in Figure 2.13.

Apart from an arbitrary time shift, and with $\omega_T = 2\pi f_T$, the nth harmonic component of $S_\tau(t)$ has the form

$$s_n(t) = (2/n\pi)|\sin(n\pi \tau/T)| \cos(n\omega_T t)$$

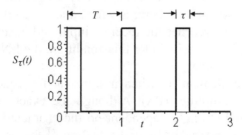

Figure 2.13   The 'chopping' function

and the component of $S\tau(t)$ has the form

$$s_2(t) = \frac{1}{\pi}\left|\sin\left(2\pi\frac{\tau}{T}\right)\right|\cos(2\ \omega_T t) \tag{2.68}$$

The amplitude of (2.68) reaches its maximal value of $1/\pi$ for $\tau/T = 1/4$. Using the latter, the 'chopping' product due to $s_2(t)$ is

$$v_{DI}(t) \cdot s_2(t)|_{\tau/T=1/4} = \tfrac{1}{\pi}\cos(2\omega_T t) \cdot V_{DI}\cos(\omega_{DI}t) \tag{2.69}$$

With $\omega_R = 2\pi f_R$, it follows from (2.64) that $\omega_{DI} = 2\omega_T - \omega_R$, and thus (2.69) yields a term $v_R(t)$ at receiver frequency of the form

$$v_R(t) = \frac{V_{DI}}{2\pi}\cos(\omega_R t) \tag{2.70}$$

The above analysis shows that the natural protection of duplex-image spur is of the order of $20\log(2\pi) \approx 16$ dB. Any further protection should come from the attenuation A_{CT} at frequency f_{DI} in the path from terminal C to terminal T, and from the attenuation A_{TR} at frequency f_R in the path from terminal T to terminal R. Thus, using the very same approach as for image rejection (section 2.6.1), we can roughly estimate a lower bound for the duplex-image rejection (*DIR*) as

$$DIR \geq 10\log(A_{CT}) + 10\log(A_{TR}) + 16 + CCR \ [dB] \tag{2.71}$$

Note that usually the attenuation A_{TR} at frequency f_R, which is required for preventing duplex desense, provides enough protection for duplex image.

2.11.2 Measurement of Duplex-Image Rejection

The measurement setup for digital modulation and using duplexer separation is according to Figure 2.14. The attenuator is required in order to prevent the transmitter power from damaging the RF signal generators.

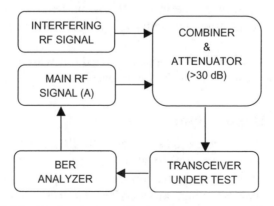

Figure 2.14   Duplex-image rejection measurement with duplexer

However, since we are interested in the ratio $S_{DI}/TSens$, there is no need to subtract the attenuator value from the level reading of the signal generators. The procedure is according to the following:

- Switch OFF signal generator B. Switch ON the transmitter. Tune signal generator A to frequency f_R, and measure the duplex sensitivity $TSens$ according to section 2.10.2.
- Increase the level of generator A by 3 dB (to the level $TS = 2TSens$).
- Switch ON signal generator B, tune it to frequency f_{DI} and increase its level S_{DI} until the receiver output deteriorates to sensitivity performance.
- Compute the duplex-image rejection using Equation (2.65).

Note that if DIR is better than blocking (section 2.8), the measurement will yield the blocking figure.

2.12 Half-Duplex Spur

The half-duplex spur is an Rx interference generated in true-duplex radios.

Denote the receiver center frequency by f_R, and the transmitter center frequency by f_T. The half-duplex frequency f_{HD} is located half-way between f_R and f_T, namely

$$f_{HD} = \tfrac{1}{2}(f_T + f_R) \tag{2.72}$$

The mechanisms causing the half-duplex spur are of higher order than the ones generating the duplex image. This is because

$$f_R = 2f_{HD} - f_T \tag{2.73}$$

which implies that a nonlinearity of at least third-order must be involved. Therefore a natural protection of the order of 35 dB or higher usually exists a priori.

Suitable mechanisms for generating the half-duplex spur are poor backward isolation of the final power device in the RF PA, or Tx power leakage to receiver input. We will not spend time analyzing this mechanism, as it is usually of secondary effect.

The measurement of half-duplex spur is carried out with a setup and procedure identical to the one described for duplex image in section 2.11.2.

2.13 Phantom-Duplex Spurs

When a portion of the Tx power as small as -10 dBm reaches the receiver LNA or mixer, it can act as a 'phantom' local oscillator signal producing IF frequency outputs whenever signals at the 'phantom image' frequency $f_{PH} = f_T \pm f_{IF}$ are present. Since the phantom image products are due to second-order nonlinearity, no natural protection exists, as compared to prime mixing products.

Just to get a feeling for a 10 W(+40 dBm) transmitter, a Tx–Rx isolation of 50 dB, although being enough to prevent receiver damage and saturation, will still generate considerable phantom spurs, unless additional attenuation for the phantom frequency is provided in the Rx front end. Along with the phantom image spur, other spurs associated with regular mixer operation, such as 'phantom half-IF', will appear too.

The measurement of the above spurs is done with the setup of Figure 2.14, using the very same procedure used for the corresponding non-duplex figures.

2.14 Conducted and Radiated Spurs

The treatment of receiver conducted and radiated spurs is identical to the transmitter case, and we defer it to Chapter 3.

3

Transmitting Systems

In this chapter, we define and explain in detail all the various transmitter specifications, and we show how to achieve, compute, and measure them.

When dealing with modern digital transmitting systems, special attention must be paid to the design of the RF power amplifier (PA). Efficiency, output power, and, above all, linearity of the PA, often constitute mandatory regulatory conditions for obtaining operating licenses in various countries. In fact, the power amplifier is the limiting factor for many, if not most transmitter specs, and often dictates the overall transmitter architecture. This is why we put special focus on it.

- High efficiency is crucial for battery lifetime in portable devices, as well as for heat dissipation.
- Output power directly determines the geographical coverage.
- Linearity is not only a limiting factor for data-rate transmission, but it is also crucial in containing the emitted RF signal within a predefined spectral mask, limiting interference to nearby receivers.

In fact, the total capacity performance of modern (and crowded) cellular systems is primarily limited by system 'carrier-to-interference' ratio (C/I), which, in turn, is heavily influenced by the linearity of the subscriber PA.

If spectral efficiency is not an issue, nonlinear amplifiers may be used, together with constant-envelope modulation schemes such as GMSK (in use with GSM systems). However, as spectral efficiency, management, and coexistence become critical, variable-envelope multilevel and multicarrier modulation methods such as quadrature amplitude modulation (QAM) combined with orthogonal frequency division multiplexing (OFDM) need to be employed.

The introduction of such high-density modulation schemes, which has been made possible by the availability of modern high-speed signal-processing devices, led to a breakthrough in spectral efficiency in harsh multipath conditions. Their application, however, requires implementation of PA linearization methods that often make use sophisticated algorithms involving the whole transmitting chain. In

Wireless Transceiver Design   Ariel Luzzatto and Gadi Shirazi
© 2007 John Wiley & Sons, Ltd

turn, the effectiveness of linearization methods strongly depends on the 'starting-point' performance of the PA itself.

Linearity, power, and efficiency are conflicting requirements, and sometimes a satisfactory trade-off between them is hard to achieve. For instance, 'class A' power amplifiers have good linearity (Clarke and Hess, 1978), but drive large DC bias current independently from the RF power supplied to the load, thus they have a very poor energy efficiency, especially when the amplified signal has high peak-to-average power ratio (PAPR). Thus many linear amplifiers make use of 'class AB' devices, whose linearity is not as good. To make things worse, the 'bare' class AB device exhibits too much distortion for many broadband applications.

In the discussion that follows we analyze the important concepts and specifications in PA design, and highlight several techniques that have been developed to deliver strong RF power while maintaining outstanding linear characteristics of the overall transmitting chain.

3.1 Peak-to-Average Power Ratio (PAPR)

As pointed out before, the demand to boost more data in a given bandwidth, improve spectrum efficiency, and still maintain the ability to deal with multipath delay spread in wireless systems, has introduced the use of sophisticated modulation schemes employing dense-constellation mapping. Unfortunately these schemes generate signals whose envelopes exhibit high peak-to-average power ratios.

For example for an 801.11a WLAN system, which uses a combined OFDM/ QAM modulation scheme with 52 carriers, the PAPR is ~10 dB, and even the low-speed narrowband TETRA (TErrestrial TRunked rAdio) standard, which uses a $\pi/4$ DQPSK modulation has PAPR ~2 dB.

Since the peak power is limited by the saturation level of the PA, it follows that the higher the PAPR, the lower the average power, and thus the shorter the transmission range.

Denote by P_{Peak} the highest possible value of the instantaneous transmitted power that may occur in a given system, and by P_{Avg} the value obtained by averaging the transmitted power over a long (ideally infinite) period of time. The PAPR is defined as the ratio

$$PAPR = \frac{P_{Peak}}{P_{Avg}} \tag{3.1}$$

and is usually expressed in dB.

3.1.1 Computation of PAPR for Quasi-Static RF Signals

For the sake of simplicity we use the following definition: a quasi-static RF signal is a modulated carrier whose center frequency is much higher than the bandwidth of the *modulating* (baseband) signal.

For practical purposes, in our context, this definition is equivalent to the definition of 'narrowband', which is somewhat more restrictive, since it implies that the center frequency is much higher than the bandwidth (centered at carrier frequency) of the *modulated* signal.

Except for a very few special schemes (not treated in this book), such as ultra-wideband (UWB) modulation, the majority of RF signals, including those referred to as 'broadband', are of the quasi-static type.

Consider the generalized (real-valued) RF signal of the form described in section 1.2.1.2:

$$S(t) = v(t) \cos[\omega t + \theta(t)], \quad -\infty < v(t) < \infty, \ \theta(t) \in [-\tfrac{\pi}{2}, \tfrac{\pi}{2}) \qquad (3.2)$$

Denoting by B_v and B_θ the bandwidth of $v(t)$ and $\theta(t)$ respectively, and denoting by B the largest between the two, the quasi-static condition implies

$$\omega \gg 2\pi B, \ B = max\{B_v, B_\theta\} \qquad (3.3)$$

The average power of $S(t)$ developed across a 1 Ω resistor in the time interval $[t_0, t_0 + T]$ is given by:

$$P_{Avg} = \frac{1}{T} \int_{t_0}^{t_0+T} S^2(t)\,dt = \frac{1}{T} \int_{t_0}^{t_0+T} v^2(t)\cos^2[\omega t + \theta(t)]\,dt$$

$$= \frac{1}{2T} \left\{ \int_{t_0}^{t_0+T} v^2(t)\,dt + \int_{t_0}^{t_0+T} v^2(t)\cos[2\omega t + 2\theta(t)]\,dt \right\} \qquad (3.4)$$

If $v(t)$ is a continuous function, and we take $\omega T \gg 1$, then there is a very general result, known as the 'Riemann–Lebesgue lemma' (Bender and Orszag, 1978), which proves that the last integral in Equation (3.4) may be neglected.

In the quasi-static case, if we take $T \geq 1/B$ then it follows from Equation (3.3) that $\omega T \geq \omega/B \gg 1$ and thus we may approximate Equation (3.4) to the form

$$P_{Avg} \approx \frac{1}{2T} \int_{t_0}^{t_0+T} v^2(t)\,dt \approx \lim_{T \to \infty} \frac{1}{2T} \int_{t_0}^{t_0+T} v^2(t)\,dt \qquad (3.5)$$

The explanation of the Riemann–Lebesgue lemma is the following: if a slowly varying function $x(t)$ is multiplied by a fast-oscillating sinusoidal function $\sin(\omega t)$, the value of $x(t)$ is nearly constant during one period of oscillation. As a result, integrating $x(t)\sin(\omega t)$ over one oscillation period yields nearly zero. It follows that the magnitude of the last integral in Equation (3.4) is at most equal to the value obtained integrating over half a cycle, thus, if the period of the sinusoid becomes very short, the value of the integral tends to zero.

In the quasi-static settings, the 'instantaneous' power at the instant t_p is defined as the average power in a time interval $[t_p, t_p + \tau]$ such that $2\pi/\omega \ll \tau \ll 1/B$. In this time interval, $v(t)$ and $\theta(t)$ are roughly constant, and $S(t)$ is nearly equal to a pure sinusoid, namely

$$S(t) = v(t_p) \cos[\omega t + \theta(t_p)]$$

$$t \in [t_p, t_p + \tau], \{t_p, t_p + \tau\} \in [t_0, t_0 + T] \tag{3.6}$$

Then, the 'peak' instantaneous power P_{Peak} developed across a $1\ \Omega$ resistor in the interval $[t_0, t_0 + T]$ is the maximal average power value obtained over all the possible intervals $[t_p, t_p + \tau]$, namely

$$P_{Peak} = \max_{t_p} \frac{1}{\tau} \int_{t_p}^{t_p+\tau} S^2(t)\, dt \approx \frac{1}{\tau} \max_{t_p} \left\{ v^2(t_p) \int_{t_p}^{t_p+\tau} \cos^2[\omega t + \theta(t_p)] dt \right\}$$

$$= \max_{t_p} \left\{ \frac{1}{2} v^2(t_p) + \frac{1}{2\tau} v^2(t_p) \int_{t_p}^{t_p+\tau} \cos[2\omega t + 2\theta(t)] dt \right\} \tag{3.7}$$

Since $2\pi/\omega \ll \tau$ then $\omega\tau \gg 1$, and by the Riemann–Lebesgue lemma it follows that the last integral is negligible, thus, we may approximate (3.7) to the form

$$P_{Peak} \approx \frac{1}{2} \max_{t \in [t_0, t_0+T]} v^2(t) \tag{3.8}$$

Finally, using (3.5) and (3.8), Equation (3.1) yields

$$PAPR = \frac{P_{Peak}}{P_{Avg}} = T \max_{t \in [t_0, t_0+T]} \{v^2(t)\} \bigg/ \int_{t_0}^{t_0+T} v(t)^2\, dt \tag{3.9}$$

Using a somewhat more mathematical language, we can say that it is natural to define the PAPR in terms of the norms of the signal, namely, with the infinity norm and the 2-norm given by

$$\|v\|_2 = \left(\int_{t_0}^{t_0+T} v(t)^2\, dt \right)^{\frac{1}{2}}, \quad \|v\|_\infty = \sup_{t \in [t_0, t_0+T]} |v(t)| \tag{3.10}$$

we get the meaningful result

$$PAPR = T(\|v\|_\infty / \|v\|_2)^2 \tag{3.11}$$

From now on we implicitly assume that all our signals are quasi-static.

Example: Consider a signal $S(t)$ consisting of the sum of N unmodulated carriers $\{S_n(t)\}$, each one of the form

$$S_n(t) = V_n \cos(\omega_n t + \theta_n), \quad |V_n| > 0, n = 1, 2, \ldots, N \qquad (3.12)$$

Assume that the frequencies ω_n are high enough and spaced apart enough so that, over the periods of interest T and τ as described before, they satisfy

$$\omega_n \tau \gg 1, \quad |\omega_n - \omega_m|\tau \gg 1, \quad n \neq m \qquad (3.13)$$

The composite signal has the form

$$S(t) = \sum_{n=1}^{N} S_n(t) = \sum_{n=1}^{N} V_n \cos(\omega_n t + \theta_n) \qquad (3.14)$$

so the average power of the composite signal is given by

$$P_{Avg} = \frac{1}{T} \int_{t_0}^{t_0+T} \left[\sum_{n=1}^{N} S_n(t) \right]^2 dt$$

$$= \frac{1}{T} \left[\int_{t_0}^{t_0+T} \sum_{n=1}^{N} S_n^2(t)\, dt + 2 \int_{t_0}^{t_0+T} \sum_{n \neq m} S_n(t) \cdot S_m(t)\, dt \right] \qquad (3.15)$$

and its peak power is

$$P_{Peak} = \max_{t_p} \frac{1}{\tau} \int_{t_p}^{t_p+\tau} \left[\sum_{n=1}^{N} S_n(t) \right]^2 dt \qquad (3.16)$$

We leave it to the reader to show that using Equation (3.13) and with the help of the Riemann–Lebesgue lemma, Equation (3.15) can be approximated to the form

$$P_{Avg} = \frac{1}{2} \sum_{n=1}^{N} V_n^2 \qquad (3.17)$$

and (3.16) is bounded by

$$P_{Peak} \leq \frac{1}{2} \left[\sum_{n=1}^{N} V_n \right]^2 \qquad (3.18)$$

If the carriers have all the same value, namely $V_n = V, n = 1, 2, \ldots N$, then the peak-to-average power ratio is bounded by

$$PAPR \leq N \qquad (3.19)$$

Note: The result above does not provide any information about the behavior of the envelope over time, and how often the peak power occurs.

In specific scenarios, it is possible to compute a probability distribution function (PDF) of the envelope, providing information on the probability of appearance of high peaks. Based on the PDF, it is possible to design effective 'clipping algorithms'.

We can guess that, in the case where high peaks appear with very low probability, they can be clipped off (at some level) generating moderate equivalent noise power, and the total degradation of the system will be tolerable, while the PAPR will be effectively reduced. For instance, in an 801.11a WLAN system using 52 carriers with BPSK modulation, the upper bound of PAPR is $10 \, log \, (52) \approx 17$ dB.

Nevertheless, if we use a PA capable of handling a PAPR of only 10 dB before clipping, it turns out that the performance of the system is still fair. The reason is that peaks of more than 10 dB above average appear in less than 1% of the time, so they can be 'chopped off' with minor consequences. A thorough analysis of envelope statistics and clipping strategy is beyond the scope of this book, and we refer the interested reader to the extensive literature available on this subject (van Nee and de Wild, 1998). Since, as pointed out earlier, the average transmitted power is the figure that ultimately determines the system performance, it is in the best interest to keep the PAPR as low as possible. Several techniques have been used to perform this task:

- The simplest method employs a 'hard-clipping' mechanism: every instantaneous peak above a predefined level is abruptly chopped off. If the occurrence of peaks is rare, the chopping introduces only a small distortion in the transmitted signal. This small distortion, does not degrade appreciably the bit error rate (BER) when detecting the clipped signal at the corresponding receiver, but the equivalent noise arising from the distortion generates an interference known as 'adjacent coupled power ratio' (ACPR), which dramatically increases the level of spurious energy spread out into the adjacent channels. At short physical distance from the transmitter, if this spurious energy, detected by nearby receivers, becomes larger than the thermal noise, it impairs their sensitivity and produces serious system performance degradation in noise-limited scenarios. For this reason hard clipping is not widely used. Instead several algorithms have been designed, which perform a gradual 'soft clipping' of the signal at the baseband stage using signal-processing techniques.
- Using differential modulation schemes such as $\pi/4$-DQPSK, provides an effective way to reduce PAPR. Here the idea is to encode the bit stream so as to avoid near-zero crossing in the complex plane path during the transition from one symbol to another, thus reducing the dynamic range of the baseband signal.
- Baseband 'peak windowing' is another popular approach. Here every peak above some predefined level is multiply by a smoothing window such as

Figure 3.1 Sensitivity measurement with digital modulation

Gaussian, Hamming, Kaiser, etc. Peak windowing provides a reasonable improvement in PAPR at the cost of a slight reduction in BER and some inherent degradation in ACPR.
- Scramblers and codes, such as the popular Gray code, are also used at symbol level with the purpose of reducing the probability of occurrence of symbol sequences that produce high peaks.

3.1.2 Measurement of PAPR

A lab setup for measuring PAPR is described in Figure 3.1. Many modern power meters have the ability to measure the peak power, the average power and the cumulative distribution function (CDF) of RF signals generated by complex baseband modulation. In order to obtain adequate measurement accuracy and protect the meter from damage the following steps are essential:

- Calibrate the power meter using a signal with a modulation scheme similar to the one to be tested, and with known characteristics.
- Verify carefully the maximum peak and average power that the power sensor is capable of handling. Set the attenuator accordingly.
  - If the attenuation level is too small, the power sensor may suffer damage.
  - If the attenuation level is too high, the measurable dynamic range may become insufficient.
- After the attenuator is set, record, at the relevant frequency band, the insertion loss of the whole combination including cables, connectors and attenuator.
- The recorded insertion loss should be the figure used to calibrate the reference level of the power meter.
  From the results of peak and average power measurements, the PAPR is computed using Equation (3.1).

3.2 Effects of Nonlinearity in RF Power Amplifiers

All RF power amplifiers exhibit some nonlinear behavior, which results in a two-fold effect on a narrowband RF signal of the form (3.2):

- Distortion of the envelope $A(t)$ due to distortion of the instantaneous signal $S(t)$, referred to as 'AM to AM conversion'.
- Distortion of the phase $\theta(t)$ due to distortion of the instantaneous signal $S(t)$, referred to as 'AM to PM conversion'.

In digital modulation schemes, such distortion critically affects symbol error rate and generates spectral spread out of the transmitted signal, causing interference to nearby channels.

The mechanisms impairing PA linearity are fairly well understood for the amplitude, but are complex and difficult to analyze as far as the phase is concerned, and both mechanisms depend on the specific PA technology. Nevertheless, as we show next, we are able to gain some insight regarding the form of their simultaneous mathematical representation.

3.2.1 Analytic Models for PA Nonlinearity

We begin with a simple and analytically convenient approach described in Heiter (1973). The input to the RF PA consists of the narrowband signal $S(t)$ in Equation (3.2). Since the modulation of $S(t)$ is quasi-static as compared to the center frequency ω, for the purpose of harmonic analysis, while the instantaneous phase ωt plays a role, $v(t)$ and $\theta(t)$ are considered to be constant.

The output $S_{out}(t)$ of the nonlinear PA can be approximated by the modified McLaurin series expansion in (3.20), which includes order-dependent time delays. In other words we assume that the nonlinearity of order n occurs after the signal experiences a delay τ_n.

$$S_{out}(t) = \sum_{n=1}^{\infty} a_n (S(t - \tau_n))^n$$

$$= \sum_{n=1}^{\infty} a_n (v \cos[\omega t + \theta - \omega \tau_n])^n, \quad a_1 = 1 \tag{3.20}$$

Setting $\phi_n = \omega \tau_n$ and taking ϕ_1 as the reference phase, we may arbitrarily set $\phi_1 \equiv 0$, so that (3.20) takes the form

$$S_{out}(t) = \sum_{n=1}^{\infty} a_n (v \cos[\omega t + \theta + \phi_n])^n, \quad \phi_1 = 0, \quad a_1 = 1 \tag{3.21}$$

Now, since all the harmonics and subharmonics of $S_{out}(t)$ are filtered out before reaching the antenna, the only products of interest in (3.21) are those that result in signals near ω, which we refer to as the 'in-band' components.

To find them out, let us rewrite $\cos^n[\omega t + \theta + \phi_n]$ in Euler form and use the binomial expansion

$$(x + y)^n = \sum_{k=0}^{n} \binom{n}{k} x^k y^{n-k} \tag{3.22}$$

getting

$$S_{out}(t) = \sum_{n=1}^{\infty} a_n \left(\frac{v}{2}\right)^n (e^{j(\omega t + \theta + \phi_n)} + e^{-j(\omega t + \theta + \phi_n)})^n$$

$$= \sum_{n=1}^{\infty} a_n \left(\frac{v}{2}\right)^n \sum_{k=0}^{n} \binom{n}{k} e^{j(\omega t + \theta + \phi_n)k} e^{-j(\omega t + \theta + \phi_n)(n-k)} \quad (3.23)$$

Equation (3.23) yields in-band components only if $k - (n - k) = \pm 1$.

It follows that only integers such that $n \pm 1 = 2k$ are of interest, thus n must be odd

$$n = 2m + 1, \quad k = \frac{n \pm 1}{2} = \{m, m + 1\}, \quad m = 0, 1, 2 \ldots \quad (3.24)$$

The in-band signal $S(t)$ arising from (3.23) takes the form

$$S(t) = \sum_{m=0}^{\infty} a_{2m+1} \left(\frac{v}{2}\right)^{2m+1} \left[\binom{2m + 1}{m} e^{-j(\omega t + \theta + \phi_{2m+1})} + \binom{2m + 1}{m + 1} e^{j(\omega t + \theta + \phi_{2m+1})} \right]$$

$$= v \cdot \sum_{m=0}^{\infty} a_{2m+1} \left(\frac{v}{2}\right)^{2m} \binom{2m + 1}{m} \cos(\omega t + \theta + \phi_{2m+1})$$

$$= v \cdot \sum_{m=0}^{\infty} c_m v^{2m} \cos(\omega t + \theta + \phi_{2m+1})$$

$$= \left[v \cdot \sum_{m=0}^{\infty} c_m \cos(\phi_{2m+1}) v^{2m} \right] \cos(\omega t + \theta)$$

$$- \left[v \cdot \sum_{m=0}^{\infty} c_m \sin(\phi_{2m+1}) v^{2m} \right] \sin(\omega t + \theta)$$

$$= V(v) \cos(\omega t + \theta + \tilde{\phi}(v)) \quad (3.25)$$

where we used

$$c_m = 2^{-2m} a_{2m+1} \binom{2m + 1}{m}, \quad \binom{2m + 1}{m} = \binom{2m + 1}{m + 1} \quad (3.26)$$

and we note that

$$c_0 = a_1 \gg 1 \quad (3.27)$$

is the 'linear voltage gain' of the amplifier. Using the Stirling formula (Abramovitch and Stegun, 1970) for large n

$$n! \approx \sqrt{2\pi n} \, n^n e^{-n}, \quad n \to \infty \quad (3.28)$$

it is easy to show that

$$2^{-2m} \binom{2m + 1}{m} = 2^{-2m} \frac{(2m)!(2m + 1)}{(m!)^2 (m + 1)} \approx \frac{2}{\sqrt{\pi m}}, \quad m \to \infty \quad (3.29)$$

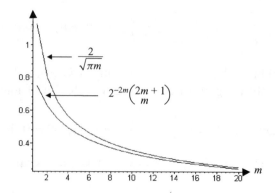

Figure 3.2   Approximation using Stirling formula

and therefore

$$c_m \approx \frac{2}{\sqrt{\pi m}} a_{2m+1}, \quad m \to \infty \tag{3.30}$$

Figure 3.2 shows the approximation in (3.29) as compared to the exact value.

If the power amplifier device is 'memory-less', namely $\phi_{2m+1} = 0, \forall m$, then there is no phase distortion and the distorted output amplitude $V(v)$ has the form

$$V(v) = v \cdot \sum_{m=0}^{\infty} c_m v^{2m} = v \cdot f(v^2) \tag{3.31}$$

where $f(v^2)$ denotes some function of v^2 such that $v \to 0$ implies $f(v^2) \to$ constant.

If the device is 'nearly-memory-less', meaning that the angles $\{\phi_{2m+1}\}$ are small (which is usually the case), we may approximate (3.25) to the form

$$S_{IB}(t) \approx \left(v \cdot \sum_{m=0}^{\infty} c_m v^{2m} \right) cos(\omega t + \theta) - \left(v \cdot \sum_{m=0}^{\infty} c_m \phi_{2m+1} v^{2m} \right) sin(\omega t + \theta) \tag{3.32}$$

We note that (3.32) is the quadrature representation (section 1.2.1.2) of a signal approximately of the amplitude of (3.31), and with an added phase $\varphi(v)$ of

$$\varphi(v) \approx tg^{-1} \left(\sum_{m=0}^{\infty} c_m \phi_{2m+1} v^{2m} \middle/ \sum_{m=0}^{\infty} c_m v^{2m} \right)$$

$$\approx \sum_{m=0}^{\infty} c_m \phi_{2m+1} v^{2m} \middle/ \sum_{m=0}^{\infty} c_m v^{2m} = g(v^2) \tag{3.33}$$

where $g(v^2)$ denotes some function of v^2 such that $g(v^2) \to 0$ when $v \to 0$.

We see that both output amplitude distortion and output phase distortion are functions of the input amplitude only.

Much effort has been invested in fitting analytical models to experimentally measured data, which lead to several widely accepted models. One of the most useful, due to its simplicity and accuracy, is the 'Saleh model' (Saleh, 1981) which exploits the mathematical architecture of (3.31) and (3.33) as follows

$$V(v(t)) = c_0 v_{sat}^2 \frac{2v(t)}{v^2(t) + v_{sat}^2}$$

$$\varphi(v(t)) = \frac{\pi}{6} \frac{2v^2(t)}{v^2(t) + v_{sat}^2} \tag{3.34}$$

Here V and φ are the compressed output amplitude and the added phase, and $v_{sat} > 0$ is the 'input saturation voltage', a rather loose definition usually determined by actual measurements. We will define v_{sat} in a reasonable and convenient way later on.

It is customary to write (3.34) in the normalized form

$$r(t) = \frac{v(t)}{v_{sat}}$$

$$\tilde{V}(r(t)) \equiv \frac{V(v(t))}{c_0 v_{sat}} = \frac{2r(t)}{r^2(t) + 1}$$

$$\tilde{\phi}(r(t)) \equiv \varphi(v(t)) = \frac{\pi}{6} \frac{2r^2(t)}{r^2(t) + 1} \tag{3.35}$$

Figure 3.3 shows a plot of Equation (3.35) with $c_0 v_{sat} = 1$.

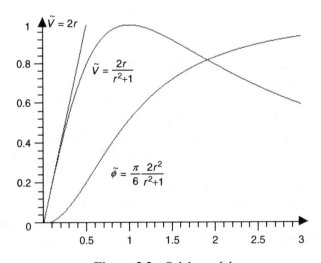

Figure 3.3   Saleh model

Note that Equation (3.35) implies that

$$|\tilde{V}(r(t))| \leq 1 \tag{3.36}$$

A more generalized form of the Saleh model uses the two-parameter formula

$$\tilde{V}(r) = \frac{\alpha_a r}{\beta_a r^2 + 1}$$

$$\tilde{\phi}(r) = \frac{\alpha_\varphi r^2}{\beta_\varphi r^2 + 1} \tag{3.37}$$

where $\alpha_a, \beta_a, \alpha_\varphi, \beta_\varphi$ are parameters to be determined by measurement.

3.2.2 Effects of PA Nonlinearity on Digital Modulation

A great deal of work has been done in analyzing the effects of transmitter distortion on the constellation of digital modulated signals. The simple normalized Saleh model is very useful for the qualitative analysis of nonlinear phenomena in digital transmission.

Many recent works (Lyman and Wang, 2002; Yu and Ibnkahla, 2005; Feng and Yu, 2004) still perform their analysis by applying the simple normalized model in (3.35) to 16-QAM (section 1.2.1.2), which is the simplest modulation scheme that clearly illustrates the phenomena generated by both amplitude and phase distortion. We stick with this approach for the analysis that follows, with the results shown in Figure 3.4.

With reference to Equation (3.2), denote by v_n and θ_n the amplitude and the phase of the symbol transmitted at time t_n, and denote by $r_n = v_n/v_{sat}$ the corresponding normalized amplitude in the normalized Saleh model.

The symbol $\tilde{V}_n$ transmitted at time t_n is the vector

$$\tilde{V}_n = \tilde{V}(r_n)e^{j[\theta_n + \tilde{\phi}(r_n)]} \tag{3.38}$$

Equation (3.35) shows that, according to the value of r_n, the symbol will be both compressed and rotated.

It follows that symbols with larger amplitude will suffer more distortion than those that are closer to the center of the constellation.

Figure 3.4 shows the effect of amplitude and phase distortion computed from (3.38), for a 16-QAM constellation for $v_{sat} = 100, 10, 3, 1.5$ and with $v_n \in \{1, \sqrt{2}, 3, \sqrt{18}\}$.

Note that for $v_{sat} = 100$, corresponding to $r_n \leq (18)^{1/2}/100 \approx 0.04$, the constellation is undistorted, while for $v_{sat} = 3$ for which $r_n \leq (18)^{1/2}/3 \approx 1.4$, the outer symbols are distorted, but the inner ones are almost unchanged. For $v_{sat} = 1.5$ the constellation is so badly distorted that no symbol can be recognized.

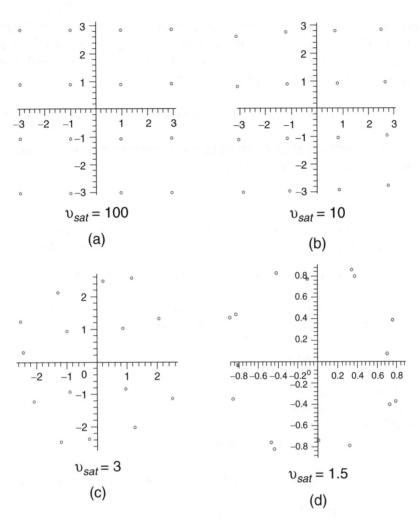

Figure 3.4 16-QAM constellation distortion using Saleh model

3.2.3 Effects of PA Nonlinearity on Spectral Shape

The effects of PA nonlinearity on the spectral shape of the transmitted signal are difficult to understand and analyze, and even tricky to simulate:

- The 'transfer function' of a nonlinear amplifier is strongly dependent on the input signal. Thus a generalized transfer function cannot be found, and each specific case must be analyzed ad hoc.
- PA nonlinearity results in bandwidth spread out: the output signal has wider bandwidth compared to the input. This effect is often referred to as 'spectral regrowth'.

- Special care must be taken with digital sampling. Because of spectral regrowth, the required sampling rate for the output signal is higher than for the input.
- The input signal must be oversampled based on an a priori (non-straightforward) estimation of the spectral shape at the output.

To understand the spectral regrowth phenomenon consider the following:

(I) Given two functions of time $g_1(t)$ and $g_2(t)$, and with $*$ denoting convolution, the Fourier transform has the property

$$F[g_1(t) \cdot g_2(t)](f) = (F[g_1] * F[g_2])(f) \qquad (3.39)$$

For any function $g(t)$, denote by $g(*)^k g$ the k-fold convolution of $g(t)$ with itself. Then, if $g_1(t) = g_2(t) = g(t)$, (3.39) can be generalized to the form

$$F[g^m](f) = (F[g](*)^{m-1} F[g])(f), \quad m = 1, 2, 3 \ldots \qquad (3.40)$$

namely, the Fourier transform of $[g(t)]^m$ can be computed by convolving $(m - 1)$ times the Fourier transform of $g(t)$ with itself, and we formally define

$$(F[g](*)^{-1} F[g])(f) = F[v^0](f) = \delta(f) \qquad (3.41)$$

(II) We denote by $supp[g]$ (the support of g), the range $[t_1, t_2], t_1 < t_2$ such that if t is outside that range, then $g(t) \equiv 0$, but $g(t_1) \neq 0$ and $g(t_2) \neq 0$. We express this in the form

$$t \not\subset supp[g] \Rightarrow g(t) = 0 \qquad (3.42)$$

(III) It is straightforward to prove that the k-fold self-convolution satisfies

$$supp[g] = [t_1, t_2] \Rightarrow supp[g(*)^k g] = [(k + 1)t_1, (k + 1)t_2] \qquad (3.43)$$

and therefore it follows from (3.40) that if $supp[F[v]] = [-B/2, B/2]$

$$supp[F[v^m]] = supp[(F[v](*)^{m-1} F[v])] = [-m\tfrac{B}{2}, m\tfrac{B}{2}], m \geq 1 \qquad (3.44)$$

meaning that, if $v(t)$ is bandlimited with bandwidth B, then $[v(t)]^m$ is bandlimited with bandwidth mB. Therefore, for even m, namely $m = 2k, k = 1, 2, \ldots$

$$supp[F[v^{2k}]] = [-kB, kB], k \geq 0, 1, 2 \ldots \qquad (3.45)$$

Thus, although the output spectrum decays fast enough to guarantee convergence, it is not necessarily bandlimited.

If the complex amplitude $a(t)$ of the input signal, is

$$a(t) = v(t)e^{j\theta(t)}, \quad |v(t)| \ll v_{sat}, \quad \theta(t) \in [-\tfrac{\pi}{2}, \tfrac{\pi}{2}) \qquad (3.46)$$

assuming, for the sake of simplicity that the distortion is memoryless, and there are only M intermodulation products, the output complex amplitude $A(t)$ has the form

$$A(t) = V(v(t))e^{j\theta(t)} \tag{3.47}$$

where, from Equation (3.31)

$$V(v(t)) = v(t) \sum_{m=0}^{M} c_m [v(t)]^{2m} \tag{3.48}$$

and then

$$F[A](f) = F\left[ve^{j\theta} \sum_{m=0}^{M} c_m v^{2m}\right](f)$$

$$= \left(F[a] * F\left[\sum_{m=0}^{M} c_m v^{2m}\right]\right)(f)$$

$$= \left(F[a] * \sum_{m=0}^{M} c_m (F[v](*)^{2m-1} F[v])\right)(f) \tag{3.49}$$

Assuming that the undistorted signal $a(t)$ has bandwidth $\pm B/2$ and the amplitude signal $v(t)$ has bandwidth $\pm Bv/2$, it follows from (3.43) and (3.45), that the support of $F[A]$ in (3.49) is

$$supp(F[A]) = [-(B/2 + MBv), (B/2 + MBv)] \tag{3.50}$$

If the signal (3.46) has only phase modulation, then v is constant, and therefore $F[v]$ is a Dirac function, from which it follows, as expected, that (3.49) yields no spectral regrowth.

Let us try to gain a feeling of the severity of the spectral regrowth of $a(t)$ by means of an example. From (3.43) and (3.46) it follows that

$$supp[F[v](f)] \subseteq supp[F[a](f)] \tag{3.51}$$

Therefore, if we assume that there is no phase modulation, this is a worst-case situation. To make things simple, let us take a signal consisting of one single small impulse bandlimited to $\pm 1/2$, namely

$$a(t) = v(t), \quad v(t) = \frac{sin(\pi t)}{\pi t} v_{sat}\varepsilon, \quad \varepsilon = \frac{\|v(t)\|_\infty}{v_{sat}} < 1 \tag{3.52}$$

The value $1/\varepsilon$ is denoted in the literature by the name 'back-off'. Before we proceed further, let us adopt a more formal definition of v_{sat}.

Assuming that the third-order product is dominant (as usually it yields much larger outputs than higher orders for relatively small signals), with v fixed, by (3.31) the output amplitude is

$$V(v) \approx c_0 v + c_1 v^3 \tag{3.53}$$

Because of the compression, when the voltage increases, the output amplitude will be less than the amplitude that would result from the linear gain c_0 alone. Note that the existence of compression implies that c_0 and c_1 have opposite signs.

We define v_{sat} as the input voltage for which the output voltage is 1 dB less than it would result from the linear gain alone. From (3.53)

$$20 \log_{10} \frac{V(v_{sat})}{c_0 v_{sat}} = 20 \log_{10} \left(1 + \frac{c_1}{c_0} v_{sat}^2\right)$$

$$= 20 \log_{10} \left(1 - \left|\frac{c_1}{c_0}\right| v_{sat}^2\right) = -1 \tag{3.54}$$

and recalling that c_0 and c_1 have opposite signs we get

$$v_{sat} \approx 0.33 \left|\frac{c_0}{c_1}\right|^{\frac{1}{2}} \tag{3.55}$$

This definition of v_{sat} corresponds to the input value known as the '1 dB compression point' of the amplifier. The Fourier transform of $v(t)$ yields

$$F[v] = \chi(f) v_{sat} \varepsilon, \quad \chi(f) = \begin{cases} 1, |f| \leq \dfrac{1}{2} \\ 0, |f| > \dfrac{1}{2} \end{cases} \tag{3.56}$$

and Equation (3.49) becomes

$$F[A](f) = \sum_{m=0}^{M} c_m (F[v](*)^{2m} F[v])(f)$$

$$= \sum_{m=0}^{M} c_m v_{sat}^{2m+1} \varepsilon^{2m+1} (\chi(*)^{2m} \chi)(f) \tag{3.57}$$

Note that if $c_m = (-1)^m |c_0|$, then (3.57) corresponds to the Saleh model. Indeed, the sum (3.31) includes a convergent geometric series that leads to the form (3.35).

Taking the reasonable assumption that the output amplitude generated by the linear gain is larger than the output amplitude generated by any higher distortion product, namely

$$|c_m| v_{sat}^{2m+1} \leq |c_0| v_{sat} \tag{3.58}$$

we may simplify and bound Equation (3.57) with the expression

$$|F[A](f)| \leq |c_0|v_{sat} \sum_{m=0}^{M} (\chi(*)^{2m}\chi)(f)\varepsilon^{2m+1} \tag{3.59}$$

In order to be able to plot (3.59), the only task we are left with is to analytically compute the convolution.

The $(n-1)$-fold self-convolution of $\chi(x)$ turns out to be associated with a known function named the 'cardinal B-spline of order n', denoted by $N_n(x)$, which has an analytic representation. In particular

$$(\chi(*)^{n-1}\chi)(x) = N_n\left(x + \frac{n}{2}\right), \quad n = 1, 2, 3, \ldots. \tag{3.60}$$

where

$$x \in [0, 1) \Rightarrow N_1(x) = 1, \quad x \notin [0, 1) \Rightarrow N_1(x) = 0 \tag{3.61}$$

and, for $n > 1$, $N_n(x)$, has the closed-form representation

$$N_n(x) = \frac{1}{(n-1)!} \sum_{k=0}^{n} (-1)^k \binom{n}{k} (max\{0, x - k\})^{n-1}, \quad n \geq 2 \tag{3.62}$$

which has support

$$supp[N_n(x)] = [0, n] \tag{3.63}$$

Note that $N_1(x)$ is nothing else but the 'box function' $\chi(x)$ shifted by half to the right, thus in (3.60) we centered $N_n(x)$ by shifting it to the left by half of its support. However, $N_n(x)$, $n > 1$, is by no means a trivial function. In fact $N_n(x)$ is a *spline*, namely, a piecewise polynomial function, consisting of a different polynomial on each subsegment $x \in [k, k+1), k = 0, 1, \ldots, n-1$.

The name 'B-splines' stresses the fact that they constitute a basis for the space of all piecewise polynomial functions. The name 'cardinal' refers to the fact that the subsegments have all unit length. A thorough treatment of cardinal B-splines can be found in Chui, (1995). The cardinal B-splines of order 1, 2, 3 and 4 are shown in Figure 3.5.

With the help of cardinal B-splines, and for a given input signal level such that $|v(t)| \leq v_{sat}\varepsilon$, (3.59) may be written in the convenient analytic form

$$\frac{|F[A](f)|}{|c_0|v_{sat}\varepsilon} \leq \chi(f) + \sum_{m=1}^{M} N_{2m+1}\left(f + \frac{2m+1}{2}\right)\varepsilon^{2m} \tag{3.64}$$

The plot of (3.64) in dB for various values of ε is shown in Figure 3.6.

Note that because of the nonlinear transfer function of the PA, the result of (3.64) is not trivially extendable to an arbitrary bandlimited signal, although, by

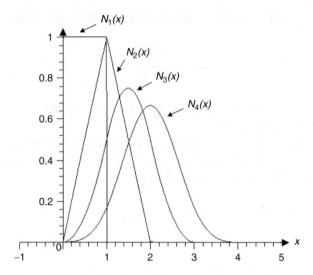

Figure 3.5   Cardinal B-splines of order 1, 2, 3 and 4

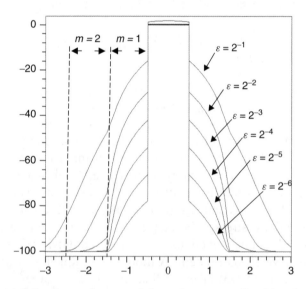

Figure 3.6   Spectral regrowth bound vs. normalized bandwidth

the sampling theorem, such a signal can be represented by a sequence of shifted pulses similar to (3.52). As a matter of fact, the bound (3.64) holds for bandlimited signals in general, as we show in the next section.

3.2.4 A Tight Bound for Spectral Regrowth

We show now that (3.64), plotted in Figure 3.6, constitutes a tight bound for the spectral regrowth of bandlimited signals.

With ε defined as before, assume that, instead of the simple signal in (3.52), we use a general bandlimited signal of the normalized form

$$v(t) = g(t)v_{sat}\varepsilon, \;\; supp(F[g]) = [-\tfrac{1}{2}, \tfrac{1}{2}], \;\; |F[g]| \leq 1 \tag{3.65}$$

Then (3.57) may be rewritten in the form

$$F[A](f) = \sum_{m=0}^{M} c_m v_{sat}^{2m+1} \varepsilon^{2m+1} (F[g](*)^{2m} F[g])(f) \tag{3.66}$$

Now, we note that, for any two functions $x(t)$ and $y(t)$

$$|x * y|(t) = \left| \int_{-\infty}^{\infty} x(\tau) y(t-\tau)\, d\tau \right| \leq \int_{-\infty}^{\infty} |x(\tau)||y(t-\tau)| d\tau = |x| * |y| \tag{3.67}$$

setting $y = x * z$ one gets from (3.67)

$$|x| * |y| \leq |x| * |x| * |z| \tag{3.68}$$

With the same approach it is easy to show by induction that the k-fold self-convolution of x satisfies

$$|x(*)^k x| \leq |x|(*)^k |x| \tag{3.69}$$

Applying the last result to (3.66) yields

$$|F[A](f)| \leq \sum_{m=0}^{M} |c_m| v_{sat}^{2m+1} \varepsilon^{2m+1} (|F[g]|(*)^{2m}|F[g]|)(f) \tag{3.70}$$

Since the convolution is carried out for non-negative functions, with $\chi(f)$ still defined by (3.56) and $|F[g]| \leq 1$, we may write, for each value of f

$$(|F[g]|(*)^{2m}|F[g]|)(f) \leq (\chi(*)^{2m}\chi)(f) \tag{3.71}$$

With the help of (3.71), equation (3.70) takes the form

$$|F[A](f)| \leq \sum_{m=0}^{M} |c_m| v_{sat}^{2m+1} \varepsilon^{2m+1} (\chi(*)^{2m}\chi)(f) \tag{3.72}$$

and using again (3.58), Equation (3.72) becomes identical to (3.59).

Thus, (3.64), which has been derived from (3.59) and plotted in Figure 3.6, is a tight bound for general bandlimited signals, where 'tight' means that one can find some function (i.e. the $sinc(x)$ pulse) that attains the bound. Therefore, a tight bound is a 'good' one (although possibly not unique).

When a precise estimate of the spectral density (section 5.3) of the output is required, the practical way to get it follows. For some time segment $T > 1/B$

- Compute as many coefficients $\{c_m\}$ as required by direct measurement on the amplifier (we show how to accomplish this task in section 3.3).
- Compute the time samples of $v(t)$ and $e^{j\theta(t)}$ at a sampling rate high enough to cover the output bandwidth of interest, including spectral regrowth.
- In the absence of signal characterization, we may take random samples with $v(t)$ normally distributed and $\theta(t)$ uniformly distributed, according to Equation (3.2).
- Compute the samples of $V(v(t))$ using Equation (3.48).
- Compute the samples of $A(t)$ using Equation (3.47).
- Compute $|F[A(t)]|^2/T$ for several time segments using FFT.
- The expected value of $|F[A(t)]|^2/T$ at each frequency constitutes the spectral density.

Example I: Figure 3.7 shows the results using the above procedure for the (normalized) values $c_0 = 1$, $c_1 = -0.05$, $c_2 = 0.003$, $c_3 = -0.0005$, and an input drive peak level of about 1 V. From (3.55) we get

$$v_{sat} \approx 0.33 \left| \frac{1}{0.05} \right|^{\frac{1}{2}} \approx 1.48$$

and since the input level is about 1 V, we set $\varepsilon = O(1)$ in (3.57).

It follows that we should expect the distortion product of order $2m + 1$ to be roughly $O(c_m)$ relative to the center of bandwidth, namely

$$m = 1 \Leftrightarrow O(c_1) = O(10^{-2}) \approx -40 \ dB$$

$$m = 2 \Leftrightarrow O(c_2) = O(10^{-3}) \approx -60 \ dB$$

$$m = 3 \Leftrightarrow O(c_3) = O(10^{-4}) \approx -80 \ dB$$

which indeed agrees with Figure 3.7. Note also that the shape of the spectral regrowth is in good agreement with the bound of Figure 3.6.

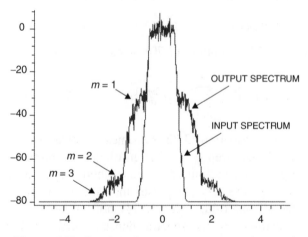

Figure 3.7   Input and output spectra superimposed

Example II: Given that the 1 dB compression point of a non-inverting amplifier is +10 dBm on 50 Ω, let us compute the amplitude portion of its extended Saleh model (the model of Equation (3.37), which we repeat here for convenience)

$$\tilde{V}(r) = \frac{V(v)}{c_0 v_{sat}} = \frac{\alpha r}{\beta r^2 + 1}, \quad r = \frac{v}{v_{sat}} \tag{3.73}$$

where +10 dBm on 50 Ω corresponds to a voltage

$$v_{sat} = v_{1dB} = \sqrt{50 \cdot 10^{-2}} \approx 0.71 V \tag{3.74}$$

Assume first, as done before, that only the third-order distortion is significant and $(\beta r^2)^2 \ll 1$ (to be verified later). Thus

$$1/(\beta r^2 + 1) \approx 1 - \beta r^2 + (\beta r^2)^2 - \dots \text{ and}$$
$$V(v) \approx c_0 v_{sat}(\alpha r - \alpha \beta r^3) + O((\beta r^2)^2) \tag{3.75}$$

Comparing (3.75) to (3.53), using (3.55) and since $|r| < 1$ we conclude that

$$\alpha \approx 1, \beta \approx \left|\frac{c_1}{c_0}\right| v_{sat}^2 = 0.109 \Rightarrow (\beta r^2)^2 < 0.01 \ll 1 \tag{3.76}$$

which verifies our previous assumption that $(\beta r^2)^2 \ll 1$, and the computed model is

$$\tilde{V}(r(t)) = \frac{r(t)}{0.109 r^2(t) + 1}, \quad r(t) = \frac{v(t)}{0.71} \tag{3.77}$$

If higher order distortion must be taken into account, then we can build an optimized model by determining α and β using the least-squares approximation on the compression points for different compression levels. For instance, assume that we have measured the 1 dB, 0.5 dB and 0.25 dB compression points getting +10 dBm, +8 dBm and +6 dBm respectively, then

$$v_{1 \ dB} = v_{sat} \approx 0.71 \Rightarrow r_{1 \ dB} = 1$$
$$v_{0.5 \ dB} = 0.56 \ V \Rightarrow r_{0.5 \ dB} = 0.56/0.71 \approx 0.79$$
$$v_{0.25 \ dB} = 0.355 \ V \Rightarrow r_{0.5 \ dB} = 0.355/0.71 = 0.5 \tag{3.78}$$

Noting that, for all values of r, the distorted output relative to the linear gain satisfies

$$\frac{V(v)}{c_0 v} = \frac{V(v)}{c_0 v_{sat} r} = \frac{\tilde{V}(r)}{r} = \frac{\alpha}{\beta r^2 + 1} \tag{3.79}$$

then we would like to simultaneously satisfy the equations

$$C_1 \equiv \frac{\tilde{V}(r_{1 \ dB})}{r_{1 \ dB}} = \frac{\alpha}{\beta(r_{1 \ dB})^2 + 1}, C_1 = 10^{-\frac{1}{20}} = 0.891$$

$$C_{0.5} \equiv \frac{\tilde{V}(r_{0.5\ dB})}{r_{0.5\ dB}} = \frac{\alpha}{\beta(r_{0.5\ dB})^2 + 1}, C_{0.5} = 10^{-\frac{0.5}{20}} = 0.944$$

$$C_{0.25} \equiv \frac{\tilde{V}(r_{0.25\ dB})}{r_{0.25\ dB}} = \frac{\alpha}{\beta(r_{0.25\ dB})^2 + 1}, C_{0.25} = 10^{-\frac{0.25}{20}} = 0.971 \quad (3.80)$$

Of course an exact solution of (3.80) is impossible in general, since there are more conditions than degrees of freedom, which yields the overdetermined system (more rows than columns) for α and β

$$\mathbf{Ax} = \mathbf{b} \Rightarrow \begin{bmatrix} 1 & -C_1(r_{1\ dB})^2 \\ 1 & -C_{0.5}(r_{0.5\ dB})^2 \\ 1 & -C_{0.25}(r_{0.25\ dB})^2 \end{bmatrix} \begin{bmatrix} \alpha \\ \beta \end{bmatrix} = \begin{bmatrix} C_1 \\ C_{0.5} \\ C_{0.25} \end{bmatrix} \quad (3.81)$$

The best we can do to 'solve' an overdetermined system $\mathbf{Ax} = \mathbf{b}$ is to find the vector $\mathbf{x}$ which yields the minimal square error (Gohberg and Goldberg, 1981), namely

$$\|\mathbf{e}\|_2 = \|\mathbf{Ax} - \mathbf{b}\|_2 \rightarrow min \quad (3.82)$$

This minimal error is achieved when the error vector is orthogonal to all the columns of the matrix (called the 'least-squares condition'), namely, by solving

$$\mathbf{A}^T(\mathbf{Ax} - \mathbf{b}) = 0 \Rightarrow \mathbf{A}^T\mathbf{Ax} = \mathbf{A}^T\mathbf{b} \quad (3.83)$$

where T indicates the transpose of the matrix. Since $\mathbf{A}^T\mathbf{A}$ is a square matrix of the smaller dimension (equal to the number of columns), in our case 2×2, and $\mathbf{A}^T\mathbf{b}$ is a vector of dimension equal to $\mathbf{A}^T\mathbf{A}$, then (3.83) is a square system, which, if $\mathbf{A}^T\mathbf{A}$ is not singular, has a unique solution. Substituting the figures of (3.80) into (3.81), and leaving the details as an exercise to the reader, the solution of our example is

$$\alpha = 1.0167, \beta = 0.1380 \quad (3.84)$$

and the model is

$$\tilde{V}(r(t)) = \frac{1.0167r(t)}{0.1380r^2(t) + 1}, \quad r(t) = \frac{v(t)}{0.71} \quad (3.85)$$

Checking the results by substituting the values of $\tilde{V}(r)/r$ at the given compression points, we get -0.985 dB, -0.578 dB and -0.226 dB, which exploit the global approximation due to least-squares. The compression in dB, $20\ log[\tilde{V}(r)/r]$, is plotted in Figure 3.8(a) for $0 < r < 1$. The normalized output $\tilde{V}(r)$ is plotted in Figure 3.8(b) for $0 < r < 5$.

3.3 Characterization of PA Nonlinearity

Let us investigate and understand several of the quantitative parameters often used by designers and manufacturers to characterize PA nonlinearity, and as a first step, let us see how to measure the coefficients $\{c_m\}$.

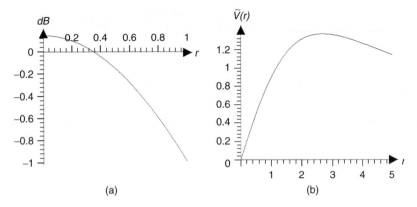

Figure 3.8   (a) Compression [dB] vs. normalized amplitude. (b) Normalized output vs. normalized amplitude

The simplest and most convenient way is to use, as the input signal to the amplifier, two sinusoids of equal and fixed amplitude v and close-by frequency namely

$$S_{in}(t) = v[cos\,(\omega - \Delta\omega)t + cos\,(\omega + \Delta\omega)t],\; \frac{\Delta\omega}{\omega} \to 0 \qquad (3.86)$$

According to (3.21) and again using the binomial expansion (3.22) for $cos^n(x)$, after a short manipulation, the resulting output signal is

$$S_{out}(t) = \sum_{n=1}^{\infty} a_n (v[cos\,(\omega - \Delta\omega)t + cos\,(\omega + \Delta\omega)t])^n$$

$$= \sum_{n=1}^{\infty} a_n (2v)^n\, cos^n(\Delta\omega t)\, cos^n(\omega t)$$

$$= \sum_{n=1}^{\infty} a_n \left(\frac{v}{2}\right)^n \sum_{k=0}^{n} \binom{n}{k} e^{j\Delta\omega t(2k-n)} \sum_{l=0}^{n} \binom{n}{l} e^{j\omega t(2l-n)} \qquad (3.87)$$

With $\Delta\omega/\omega \to 0$ and n finite, Equation (3.87) yields in-band components (in the vicinity of ω) only for $2l - n = \pm 1$, from which it follows that n must be odd. Setting $n = 2m + 1$, yields the condition

$$l = \{m, m + 1\}$$

Substituting the latter in (3.87) and using (3.26), the resulting in-band signal $S(t)$ is

$$S(t) = \sum_{m=0}^{M} a_{2m+1} \left(\frac{v}{2}\right)^{2m+1} \sum_{k=0}^{2m+1} \binom{2m+1}{k} \times e^{j\Delta\omega t(2(k-m)-1)} 2 \binom{2m+1}{m} cos\,(\omega t)$$

$$= \sum_{m=0}^{M} v^{2m+1} 2^{-2m} a_{2m+1} \binom{2m+1}{m} \times \sum_{k=0}^{2m+1} \binom{2m+1}{k} e^{j\Delta\omega t(2(k-m)-1)} \cos(\omega t)$$

$$= \sum_{m=0}^{M} c_m v^{2m+1} \sum_{k=0}^{2m+1} \binom{2m+1}{k} e^{j\Delta\omega t(2(k-m)-1)} \cos(\omega t) \tag{3.88}$$

Now, setting $k = 2m + 1 - l$, let us split the last summation in (3.88) as follows:

$$\sum_{k=0}^{2m+1} (\cdot) = \sum_{k=0}^{m} \binom{2m+1}{k} e^{j\Delta\omega t(2(k-m)-1)} + \sum_{l=m}^{0} \binom{2m+1}{2m+1-l} e^{j\Delta\omega t(2(m+1-l)-1)}$$

$$= \sum_{k=0}^{m} \left[\binom{2m+1}{k} e^{j\Delta\omega t(2(k-m)-1)} \right.$$

$$\left. + \binom{2m+1}{2m+1-k} e^{-j\Delta\omega t(2(k-m)-1)} \right] \tag{3.89}$$

Using in (3.89) the property of the binomial coefficients

$$\binom{p}{q} = \binom{p}{p-q} \tag{3.90}$$

substituting the result into (3.88) and setting $u = m - k$ we finally obtain

$$S(t) = \sum_{m=0}^{M} c_m v^{2m+1} \sum_{k=0}^{m} \binom{2m+1}{k} \times 2 \cos(\Delta\omega t(2(k-m)-1)) \cos(\omega t)$$

$$= \sum_{m=0}^{M} c_m v^{2m+1} \sum_{u=0}^{m} \binom{2m+1}{m-u} \times [\cos(\omega t - (2u+1)\Delta\omega t)$$

$$+ \cos(\omega t + (2u+1)\Delta\omega t)] \tag{3.91}$$

Equation (3.91) implies that the nonlinearity of order $2k + 1$, generates sidebands at $\omega \pm (2u+1)\Delta\omega$, $u = 0, 1, 2, ..k$.

Note that c_{k-1} does not contribute at all to the kth sideband, and if v is taken to be small enough we can satisfy

$$c_m v^{2m+1} \ll c_k v^{2k+1} \binom{2k+1}{0} \Rightarrow \frac{c_m}{c_k} v^{2(m-k)} \ll \binom{2m+1}{m-k}, \quad k < m \leq M \tag{3.92}$$

Then (3.91) shows that the contribution to the kth sideband due to c_m, $m > k$ is negligible as compared to the contribution due to c_k. Thus one can estimate $|c_k|$ form the spectral picture, as if it was the only non-zero coefficient. Specifically, denoting by s_k the amplitude of the kth sideband observed, up to a constant multiplier

$$s_k \approx |c_k v^{2k+1}| \Rightarrow |c_k| \approx \frac{s_k}{|v^{2k+1}|} \tag{3.93}$$

In this case, given the linear gain c_0, the coefficients $|c_k|, k = 0, 1, 2..$, are readily computed from the spectral picture.

$$\frac{|c_{k+1}|}{|c_k|} v^2 = \frac{|c_{k+1} v^{2k+3}|}{|c_k v^{2k+1}|}$$

$$\approx \frac{\text{amplitude of } (k+2)\text{th sideband}}{\text{amplitude of } (k+1)\text{th sideband}} \tag{3.94}$$

Therefore, all we should check, in order to verify that (3.93) holds, is to verify that increasing/decreasing v does not change appreciably the ratio (3.94). In the vast majority of real-life situations, (3.94) will indeed hold, as can be seen in the following interesting example.

Example: We compare the simulated spectral shape obtained using the Saleh model presented in (3.35), with the figures obtained by direct computation of (3.94). Recall first that the model (3.35) implies that

$$c_m = (-1)^m, m = 0, 1, 2, \ldots$$

Indeed, with this choice, the sum in (3.31) is a geometric series converging to $1/(v^2 + 1)$, and (3.94) has the constant value $40 \, log(v)$ dB for all k.

We perform the simulation as follows: with $\omega = 2\pi f$, $\Delta\omega = 2\pi \Delta f$, we first compute 1024 samples of (3.86) for

$$f = 64, \Delta f = 2, v = 10^{-1}, 10^{-2}, 10^{-3}, 10^{-4}$$

We perform the FFT on each one of the sets of samples, and we plot the results in the neighborhood of $f = 64$. The results are plotted in Figure 3.9 showing that the computed values closely match the simulation, and for each value of v, the ratio between subsequent sidebands is indeed very close to $40 \, log(v)$ dB. Thus (3.91) may be approximated by (3.94).

Now, we compute the sideband amplitudes using (3.91), where, for each sideband we sum up all the contributions generated for $M = 20$, namely, taking into consideration 20 non-zero coefficients $\{c_m\}$ in (3.91). We plot the above amplitudes (in dB) connected in a solid line on the right side of the FFT plot (the left side is identical).

3.3.1 Intermodulation Distortion (IMD)

We saw that RF PA distortion is difficult to quantify, model and simulate. The 'intermodulation distortion' (IMD) is a practical parameter that provides a meaningful and quantitative measure of PA nonlinearity, it can be measured by simple

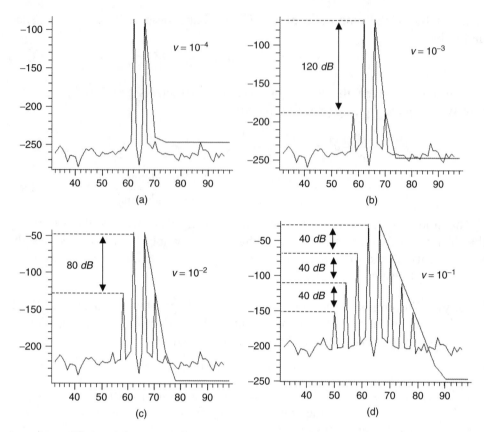

Figure 3.9 Simulated vs. computed sidebands for Saleh model

methods, and yields an 'educated guess' of the spectral spread at the various power levels. When a multi-carrier or multi-carrier 'like' (multi-narrowband) signal is injected into a nonlinear RF PA, we end up with an output that contains the 'fundamental' signal, linearly amplified, plus additional 'spurious' products. Apart from the PA characteristics, the level of the spurs depends on the average input power and on the PAPR of the input signal, so we must first define the above quantities. It is customary to use the following settings:

- The input signal $S_{in}(t)$ consists of two unmodulated carriers of equal and constant amplitude v, with symmetric frequency offset with respect to the transmit frequency f_T, but still within the transmit band

$$S_{in}(t) = v(cos\ 2\pi f_1 t + cos\ 2\pi f_2 t)$$
$$f_1 = f_T - \Delta f,\ f_2 = f_T + \Delta f \qquad (3.95)$$

- Equation (3.95) may be rewritten in the form

$$S_{in}(t) = 2v\ cos\ (2\pi\ \Delta f t)\ cos\ (2\pi f_T t) \qquad (3.96)$$

and applying (3.5) and (3.8) to (3.96), the peak and average input powers across a 1 Ω resistor are

$$P_{Peak} = 2v^2, \quad P_{Avg} = \lim_{T \to \infty} \frac{2v^2}{T} \int_{t_0}^{t_0+T} cos^2(2\pi \Delta f t) \, dt = v^2 \tag{3.97}$$

Thus $S_{in}(t)$ has a PAPR of 3 dB, and the total average power is the twice the power of one sideband.

Now, we set the average power at a level equal to the 1 dB compression point, namely $v = v_{sat}$. With v_{sat} according to (3.55), and recalling that c_0 is the linear (small-signal) voltage gain of the amplifier, the coefficients $\{c_k\}$, in (3.26), are readily found from (3.94) for $k = 1, 2, 3, \ldots$ by inspecting the output sidebands on a spectrum analyzer

$$\frac{|c_k|}{|c_0|} v_{sat}^2 \approx \frac{\text{amplitude of } (k+1)\text{th sideband}}{\text{amplitude of fundamental sideband}} \tag{3.98}$$

and then, the corresponding distortion coefficients $\{a_{2k+1}\}$ in the power series expansion (3.21) are readily found from (3.30).

The inverse of the ratio (3.98) in dB, is called the 'intermodulation distortion ratio of order N' (IMDN), $N = 3, 5, 7 \ldots$

$$IMDN = 10 \log \left(\frac{\text{power of fundamental sideband}}{\text{power of } [(N+1)/2]\text{th sideband}} \right) \tag{3.99}$$

If an input power level other than $v = v_{sat}$ is being used, it should be specified in order for IMDN to be meaningful.

It is often preferred to quantify the transmitter distortion by means of the 'N-th order input intercept point' (IPNi), a useful parameter expressed in dBm, widely used for the general characterization of nonlinear behavior in both transmitters and receivers (Chapter 6). The nice thing is that IPNi may be computed directly from IMDN, and once it is known, it is possible to compute IMDN for any arbitrary value of input power. Indeed, denoting by p (expressed in dBm) an arbitrary average power developed across a 1 Ω resistor, the relationship between the above two quantities, is given by

$$IMDN = (N - 1)(IPNi - p) \, [dB]$$
$$p = 10 \log(v^2/10^{-3}) \, [dBm] \tag{3.100}$$

Note: Denoting by $p = p_{1 \, dB}$ the power corresponding to the 1 dB compression point of a transmitter, we prove in Chapter 6 the widely used rule of thumb

$$IP3i \approx p_{1 \, dB} + 10 \, [dBm] \tag{3.101}$$

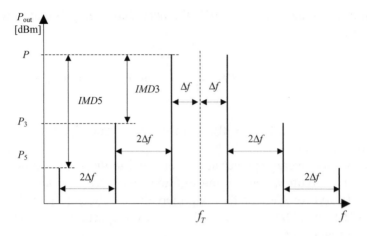

Figure 3.10 Spectral sketch of *IMD*3 and *IMD*5 products

A sketch of the spectral shape for IMD3 and IMD5 products is shown in Figure 3.10.

Example: Let us compute *IP3i* and *IP5i* from the spectral shape given in Figure 3.9 for the Saleh model. Using (3.100)

$$IPNi = p + \frac{IMDN}{N-1} \qquad (3.102)$$

Note that, since the intercept point does not depend on the drive level, it may be computed from any one of the figures. For the reasons pointed out in Chapter 6, however, it is best to perform the measurement at the smallest drive level where the corresponding IMD product becomes clearly visible.

We compute *IP3i* and *IP5i* from Figure 3.9(b), and *IP5i* from Figure 3.9(d)

- *IP3i*: from Figure 3.9(b)

$$p = 10 \log (v^2/10^{-3}) = -30 \ dBm$$

$$IMD3 = 120 \ dB$$

$$\Rightarrow IP3i = -30 + \frac{120}{2} = +30 \ dBm$$

- *IP5i*: from Figure 3.9(d)

$$p = 10 \log (10^{-2}/10^{-3}) = +10 \ dBm$$

$$IMD5 = 80 \ dB$$

$$\Rightarrow IP5i = 10 + \frac{80}{4} = +30 \ dBm$$

We could have guessed the above results in advance, since we saw that for the Saleh model, for all values of v, the power ratio between the sideband k and the sideband $k + 1$ is $-40 \log(v)$ dB, thus

$$IMDN = (N - 1)(-20 \log(v))$$

and (3.102) yields

$$p = 10 \log(v^2/10^{-3})$$

$$IMDN = -(N - 1) \cdot 20 \log(v)$$

$$\Rightarrow IPNi = 10 \log(1/10^{-3}) = +30 \ dBm$$

Not surprisingly, all the intercept points of the Saleh model in the example of section 3.3, have the identical value of +30 dBm.

3.3.1.1 Measurement of IMDN

The setup for the measurement of IMDN is described in Figure 3.11. The signal generators A and B are used to feed the two-tone fundamental sidebands to the transmitter under test. Special care should be taken to verify that the measured IMDN is indeed due to the transmitter under test.

Overlooking setup nonlinearity is a common mistake that often ends up in erroneous measurements due to masking effects of other nonlinear devices in the setup itself. Some important points are:

- The attenuators in the setup of Figure 3.11 are an absolute requirement, and must have the highest possible attenuation value (usually ≥ 10 dB each) that still allows detection of the lowest sideband of interest. This is because generators A and B, when delivering relatively high power levels, usually work without

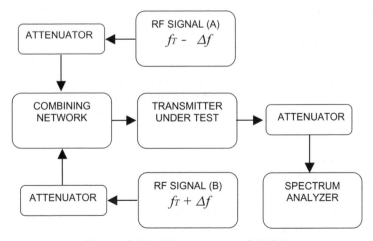

Figure 3.11   Measurement of *IMDN*

internal attenuators, and control the output RF level by controlling the driver of the output stage. If the generators are not well isolated, the mutual leakage may generate, within the output stages, parasitic intermodulation sidebands that will enter the unit under test, and will appear at the output. Since the parasitic sidebands are indistinguishable from the ones due to the unit under test, unless external attenuators are used, we may end up measuring the IMDN of the signal generators, which may be worse than the IMDN of the transmitter. This is a very common mistake that leads to wrong and puzzling results. The error may be avoided by checking the spectral picture at the output of the combining network, to verify that any parasitic IMDN reading is much better than the one being measured at the output of the transmitter.

- Another nonlinear element in the chain is the spectrum analyzer which has a limited dynamic range, so we may end up measuring the IMDN of the setup. In most cases adding an attenuator in front of the analyzer can resolve the issue. Again, the error may be avoided by checking the spectral picture while feeding the input of the spectrum analyzer with two signals of the same level expected from the transmitter, to verify that any parasitic sidebands are much lower than the ones to be measured.

3.3.2 Error Vector Magnitude (EVM)

We showed in section 1.2.1.2 that digital modulation schemes such as QAM are defined by two-dimensional constellations, in which each symbol is a phasor (a complex number) corresponding to a certain instantaneous phase and amplitude of the transmitted carrier. For a single-carrier QAM modulation, the transmitted signal consists of a sequence of carrier bursts, each of duration referred to as 'the symbol duration', whose corresponding phasors describe time-varying patterns in the complex plane.

We pointed out that symbol detection is performed by sampling the received phasors at certain equally spaced optimal sampling instants that depend, among other things, on the characteristics of the filters in the transmitter path as well as on the transmission channel (Proakis, 1983). We saw that in a non-ideal environment, the patterns of the received phasors may never cross the exact symbol position. In particular using the Saleh model, we deduced that PA nonlinearity may produce extreme a priori distortion of the transmitted constellation. The EVM is a convenient measure of the distortion in single-carrier QAM constellations and may be defined in several ways. We choose the definition which we believe is the most suited for practical use:

- Denote by $\{V_n\}$, $n = 1, 2, .., N$ a sequence of N symbols to be transmitted, each of them exactly corresponding to a QAM symbol, and define the corresponding complex vector $\mathbf{V} = [V_1, V_2, \ldots, V_N]$.
- Denote by $\{\tilde{V}_n\}$, $n = 1, 2, .., N$ the sequence of symbols detected by an ideal receiver over an ideal channel, and sampled at the proper optimal sampling instants, and denote by $\tilde{\mathbf{V}} = [\tilde{V}_1, \tilde{V}_2, \ldots, \tilde{V}_N]$ the corresponding complex vector.

- If there is no transmitter distortion the sampled phasors will be identical to the exact ones.
- Denote by $\mathbf{E} = \tilde{\mathbf{V}} - \mathbf{V}$ the error vector, then with $\|\mathbf{X}\|_2$ denoting the Euclidian norm of the vector $\mathbf{X}$, we define

$$EVM = \frac{\|\mathbf{E}\|_2}{\|\mathbf{V}\|_2} = \frac{\left(\sum_{n=1}^{N} |\tilde{V}_n - V_n|^2\right)^{1/2}}{\left(\sum_{n=1}^{N} |V_n|^2\right)^{1/2}} \tag{3.103}$$

Then EVM^2 is a measure of the ratio between the average noise power generated by all distortion effects, and the total average power of the transmitted signal.

Figure 3.12 shows an example of the exact and the distorted symbols in a 16-QAM constellation grid at the nth optimal sampling instant.

In actual digital receivers the quadrature channels are sampled separately, thus, setting

$$V_n = I_n + jQ_n$$
$$\tilde{V}_n = \tilde{I}_n + j\tilde{Q}_n \tag{3.104}$$

we get

$$EVM^2 = \frac{\sum_{n=1}^{N} [(\tilde{I}_n - I_n)^2 + (\tilde{Q}_n - Q_n)^2]}{\sum_{n=1}^{N} [I_n^2 + Q_n^2]} \tag{3.105}$$

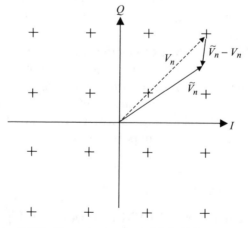

Figure 3.12   Error vector in a 16-QAM constellation

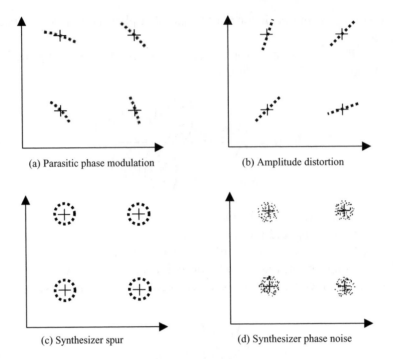

Figure 3.13 Corrupted symbols due to transmitter lineup imperfections

In other settings, the error magnitude is normalized to the magnitude of the largest symbol, so that both peak and rms *EVM* may be defined, namely

$$EVM\,|_{rms} = \frac{1}{\sqrt{N}}\frac{\|\mathbf{E}\|_2}{\|\mathbf{V}\|_\infty}, \;\; EVM\,|_{Peak} = \frac{\|\mathbf{E}\|_\infty}{\|\mathbf{V}\|_\infty} \tag{3.106}$$

where $\|\mathbf{X}\|_\infty$ denotes the magnitude of the largest component of $\mathbf{X}$.

For example, in GSM systems both rms and peak EVM are defined as 7% and 22%, respectively, while in 802.11a systems the rms value is specified as 5.6% and the peak is not defined.

In many cases, when the EVM exceeds the requirements, plotting the sequence of detected symbols superimposed in the complex I/Q plane may provide a good visual diagnosis of the source of distortion in the transmitter chain.

Figure 3.13 presents a sketch of a sequence of detected symbols for a 16-QAM constellation (only the first quadrant is shown), corrupted by various imperfections of the transmitter lineup and superimposed on the ideal constellation map (see Chapter 6 for parasitic phenomena).

3.3.2.1 Measurement of EVM

The setup for the measurement of EVM is described in Figure 3.14. It includes the transmitter, an attenuator and a vector signal analyzer (VSA).

Figure 3.14 Measurement of error vector magnitude

The VSA is an FFT-based spectrum, time, and code-domain analyzer capable of generating and detecting baseband signals for digital schemes such as QAM. The VSA provides information on signal quality such as EVM, peak and average power, ACPR (see section 3.2.7) and more.

One word of caution: care should be taken to maintain the RF input level of the VSA in the recommended range, in order to avoid compression which may result in erroneous EVM reading.

3.3.3 Adjacent Coupled Power Ratio (ACPR)

The adjacent coupled power is a measure of how much a transmitter may cause interference to nearby receivers, due to the cumulative effect of nonlinearity, phase noise, close-in spurs etc. generating spectral regrowth. No protection exists against it on the receiving side, because the interference is right on channel. This 'leakage' of the transmitter into contiguous channels is one of the worst and unpredictable 'system killers'. It may cause strong receiver desense over a wide area, sometimes leading to system crash in crowded electromagnetic environments.

The value of ACPR, expressed in dB relative to the average transmit power, is the signal power that will be detected by a receiver having a predefined bandwidth, and with center frequency outside, but close, to the transmit channel, as shown in Figure 3.15.

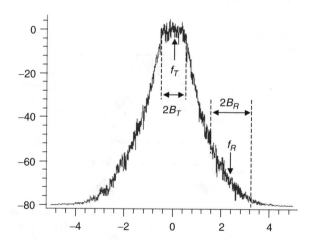

Figure 3.15 Adjacent coupled power

With f_T, $2B_T$ and f_R, $2B_R$ denoting respectively the center frequency and bandwidth of the transmitter and of the receiver, and with $S(f)$ [watt/Hz] denoting the spectral density of the transmitted signal, we define

$$ACPR = 10\,log\left\{ \int_{f_R-B_R}^{f_R+B_R} S(f)df \bigg/ \int_{f_T-B_T}^{f_T+B_T} S(f)df \right\} [dBc] \qquad (3.107)$$

Although usually B_T and B_R are identical, this is not a must.

Example: The TETRA standard is employed mainly in Europe for public safety. It uses one 25 kHz channel to transfer four voice calls. The spec for ACPR at adjacent channel is -60 dBc measured at 25 kHz off from transmit center frequency, -70 dBc at 50 kHz and at -74 dBc at 1 MHz. The power is integrated using a bandpass 'root-raised cosine' filter with $\alpha = 0.35$ and bandwidth of 18 kHz.

3.3.3.1 Measurement of ACPR

The measurement is done in a setup identical to Figure 3.14 for measuring EVM. The VSA may be replaced by a scalar spectrum analyzer.

3.3.4 Spectral Mask

The spectral mask defines a set of boundaries, usually dictated by regulatory constraints, which the transmitted signal power is not allowed to exceed. The typical structure of a spectral mask is shown in Figure 3.16.

In order to be meaningful, the mask must be defined in correspondence with specific setup settings, including integration bandwidth, averaging method, keying sequence, modulation scheme etc.

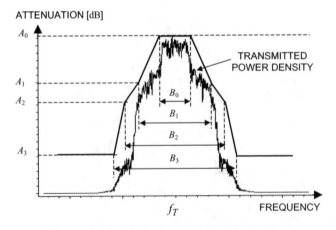

Figure 3.16   Typical spectral mask structure

Example: For the 801.11a standard in the 5 GHz frequency range with channel bandwidth of 20 MHz the values in Figure 3.16 are

- $B_0 = 18$ MHz, $A_0 = 0$ dB
- $B_1 = 22$ MHz, $A_1 = -20$ dB
- $B_2 = 40$ MHz, $A_2 = -28$ dB
- $B_3 = 60$ MHz, $A_3 = -40$ dB

3.3.4.1 Measurement of Spectral Mask

The measurement is done in a setup identical to the one used for ACPR.Care should be taken in adjusting the spectrum analyzer with the specific settings defined under the standard for which the measurement is done.

3.4 PA Efficiency

RF power amplifiers convert direct current (DC) power to RF power. This conversion process, however, is usually inefficient, in the sense that the energy driven from the DC supply is substantially greater than the RF energy delivered to the load.

The efficiency η is defined as the ratio, usually expressed in percent, between the average RF power delivered to the load, and the DC power delivered by the supply. For a finite time segment T

$$\eta = \frac{P_{Avg}}{V_{cc}I_{Avg}}, \ P_{Avg} = \frac{1}{T}\int_0^T P(t)\,dt, I_{Avg} = \frac{1}{T}\int_0^T I(t)\,dt \qquad (3.108)$$

where $P(t)$ denotes the instantaneous RF power delivered to load, V_{cc} is the fixed DC voltage, and $I(t)$ is the instantaneous current driven from the power supply, and is a function of $P(t)$.

The difference between the DC power supplied in the process and the RF power delivered to load dissipates in the form of heat in the final RF device(s) of the PA. In turn, heat dissipation dictates the device size, whose physical contact area with the heat sink must increase with dissipation, and of course dictates the size of the heat sink itself. To make things worse, heat generation comes at the expense of the working time in battery-operated portable transceivers. In this section we refer only to power amplifiers operating in the modes known as 'class AB' and 'class B', which are the ones fit for delivering high power with linear characteristics and good efficiency. For those classes, we make the following reasonable assumptions:

1. As far as we remain within the safe device operating limits, the larger the supply voltage V_{cc}, the higher the RF power that the PA can deliver to its load. This is usually true since increasing the supply voltage tends to drive the final RF device(s) out of saturation.

2. As long as V_{cc} is larger than the value required to deliver the instantaneous power, then the current driven from the DC supply is independent from V_{cc}. This is usually true since when no saturation occurs, most RF devices tend to behave like current sources.

Now, we show (by rather heuristic arguments) that signals with larger PAPR make the PA operate with worse efficiency.

It follows from assumptions 1 and 2 that if two signals S_1 and S_2 have large peak-to-average ratios, $PAPR_1$ and $PAPR_2$ respectively, with $PAPR_1 < PAPR_2$, the DC supply voltages needed for proper operation are $V_{cc1} < V_{cc2}$.

A large peak-to-average ratio means that the power peaks last only for a small percentage of the total time, and thus their contribution to the total current consumption is assumed to be negligible.

By assumption 2, the current is independent from V_{cc1} and V_{cc2}. Thus if both S_1 and S_2 have the same average power, they drive roughly the same average current I_{Avg} from the DC supply regardless of $PAPR$. It follows that the DC power delivered by the supply for S_1 is $V_{cc1}I_{Avg}$, while for S_2 is $V_{cc2}I_{Avg}$. Since $V_{cc1} < V_{cc2}$, then S_2 drives more DC energy from the power supply, while delivering the same RF energy to load.

In view of the above, as we will see later, many modern PAs use techniques called 'amplitude-tracking supply'. An amplitude-tracking supply makes use of time-variable DC supply sources, which are controlled by the instantaneous peak power level, and keep V_{cc} at the minimal value required on the fly, leading to substantial efficiency improvement.

3.5 Transmitter Transients

Most modern transceivers work in intermittent mode, periodically switching between transmit and receive state. As soon as the transmitter is keyed on, several transient phenomena occur:

- The power amplifier starts driving large current from the DC supply. This sudden current change through the power lines on the printed circuit board, causes the appearance of parasitic voltages (Chapter 6) that either develop because of the serial resistance in the printed lines, or by electromagnetic induction. These parasitic voltages may develop in series to frequency-sensitive components, such as VCO varactors, may be induced onto the VCO resonator causing injection-locking phenomena (Chapter 6), or may appear as a supply voltage ripple on gain-sensitive circuits such as modulating stages. The resulting frequency and amplitude modulation transients lead to the appearance of spurs in the transmitted spectrum. These phenomena usually appear at low repetition rate, and thus are difficult to observe and diagnose, and their correction is an extremely painful task, often requiring major hardware redesign.

- In most architectures, the synthesizer switches from receive LO frequency to transmit frequency. If RF power transmission begins before the frequency is properly stabilized, we may end up with either degradation in channel performance due to partial loss of transmitted frames, or with desense of nearby receivers due to transmit power invading adjacent channels. Fortunately, the measurement of this phenomenon is pretty straightforward (Chapter 4), and in most cases the problem may be corrected by software intervention alone.

3.5.1 Attack Time

The 'attack time' is defined as the time elapsed from transmitter key-on to the instant when the output power reaches 90% of its final value. It is usually determined by all the processes involved, including time delays generated by software. The measurement is done in repetition mode, where the key-on command serves as the trigger for the measuring equipment. We note that most of the transmit current is usually driven by the final PA device, then the RF power is roughly proportional to the DC supply current. Therefore the attack time can be roughly measured as per Figure 3.17, by just monitoring the DC supply current on a scope, and looking for the instant where it reaches 90% of its final value.

3.5.2 Frequency Shift Upon Keying

As mentioned before, when the transceiver is keyed to transmit mode, unless it has a full-duplex architecture, the synthesizer must switch from receive LO frequency to transmit frequency.

The PA control must be designed so that no RF power is allowed to build up before the synthesizer has stabilized close enough to the final frequency, so that proper transmit operation as well as regulatory requirements are met.

Note that although the synthesizer itself may have been properly designed, conducted or induced transient voltages of the type mentioned in section 3.5 may pull it out of frequency momentarily, which initiates a relock process and increases the frequency stabilization time. Therefore correct synthesizer performance does not guarantee proper keying operation, which is a system feature.

The measurement of frequency shift upon keying is done with a procedure and setup identical to the one used for measuring synthesizer lock time, and the reader is referred to Chapter 4 for its description.

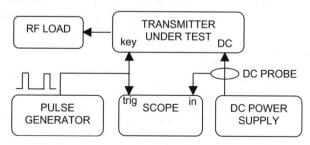

Figure 3.17   Setup for the measurement of attack time

3.6 Conducted and Radiated Emission

Unwanted transmitter emission is a key issue from regulatory, safety and system performance aspects. Its measurement, however, is difficult and labor intensive, and in some cases, as with 'specific absorption rate' (SAR) (Chapter 6), we cannot even get a rough estimate, unless very sophisticated and expensive dedicated equipment is available.

The measurements we describe here can be performed with conventional lab equipment, although, for radiated emission, the resulting figures are not much more than an 'educated guess', and the final measurements must be carried out with the help of a specialized antenna site.

The reason why we may want to bother with approximate radiation measurements is that, in a product development process, even a rough estimate of the emission performance is extremely useful. Knowledge of the order of magnitude of the radiated emission provides valuable hints regarding potential problems, and may reveal the need of an a priori layout or shielding of the printed circuit board, before we even attempt to meet the required radiation mask, thus drastically reducing the number of costly and time-consuming turnarounds with external labs.

3.6.1 Conducted Spurs

The 'conducted spurs' are the collection of all unwanted RF signals outside the designated transmit channel bandwidth, which can be observed at the antenna port during transmission, by 'notching out' the on-channel transmit power at the carrier frequency f_T, and then attaching a spectrum analyzer as shown in Figure 3.18.

Since all we need is a notch filter, a protective attenuator and a spectrum analyzer, the measurement of conducted emission is relatively straightforward. However, since many of the unwanted signals lie at unknown frequencies, it may require an extensive scanning search.

We can divide the conducted spurs in three categories

- *Harmonic spurs:* all the carrier-like (with bandwidth of the order of one to few channels) spurs that are multiples of the transmit frequency. Harmonic spurs are straightforward to measure, because their location is known a priori.
- *Non-harmonic spurs:* all the carrier-like spurs that appear at frequencies different from a multiple of the transmit frequency. They include, for instance, local oscillator leakage, synthesizer-generated spurs, clock feed-through from

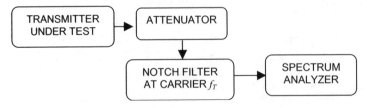

Figure 3.18   Setup for the measurement of conducted emission

digital processors etc. Non-harmonic emission, in principle, is straightforward to measure, but is hard to discover since its frequency is unknown. Since the spurs are usually embedded in wideband noise, they cannot be seen unless the spectrum analyzer bandwidth is set to narrow values, and a narrow bandwidth implies a slow and lengthy search even when done automatically.

• *Spectral bumps:* all the spurs that appear as 'hills' with bandwidth significantly wider than one channel. We should be worried about these spurs, as they often indicate the existence of potential PA instability in that frequency neighborhood. In fact, most spectral bumps are nothing more than highly amplified transmitter noise. Their existence indicates that the PA exhibits an extremely high small-signal gain in presence of large-signal operation. This high gain, is likely the result of positive feedback occurring at that specific frequency as a result of phase-shift accumulation due to parasitic resonances in the matching circuitry. If left untreated, spectral bumps often degenerate into PA oscillations over temperature, mismatch or aging.

3.6.2 Back Intermodulation

During transmission, external signals, such as carriers from powerful nearby inter-ferers, either far away or near the transmit frequency, are received by the antenna and reach the PA output. With a mechanism called 'back intermodulation', similar to the one generating the 'duplex image' and described in Chapter 2, the above signals mix with the transmit frequency and create spurs that are re-radiated from the antenna.

The back intermodulation is the re-radiated power expressed in dB below the power of the incoming interferer.

Apart from the Tx matching circuitry, almost no natural protection exists against this first-order mixing.

Fortunately, if the interfering frequency is in-band, the generated mix is off-band, and in order to generate an in-band mix, the interferer must come off-band. Thus some protection may be gained by adding a bandpass filter at the output of the PA. The setup for measuring back intermodulation is shown in Figure 3.19.

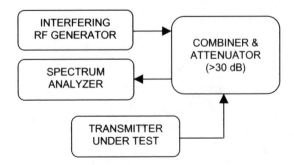

Figure 3.19   Setup for the measurement of back intermodulation

3.6.3 Radiated Spurs

Radiated spurs are the result of the transmit carrier and its harmonics, as well as various high-level local oscillators and clocks and their harmonics, radiating directly from the transceiver body or through the DC power lines. We usually know the expected radiated spurious frequencies and where to look for them. The problem here is the quantitative value.

Proper measurements require antenna sites with arrangements to prevent reflections and interferences. Here, we do not get into the various complex considerations involved in constructing such a site, but we show how we can get an approximated estimation of radiated spurs by using standard lab equipment. In order to proceed we need to refresh a few concepts.

3.6.3.1 Radiation Basics

Let us recall some basics from electromagnetic propagation and antenna theory (Balanis, 1997). The following results are frequency-dependent, and valid for a narrowband signal with some specific center frequency.

With reference to the spherical coordinates in Figure 3.20, assume that there is a radiating source S located at the origin of the system. Assume that the total radiated power is P_T [watt]. If we surround the source with an imaginary sphere of radius r, the total outward propagating power crossing the surface area of the sphere must be P_T. It follows that, if no reflection or power dissipation occur in the volume between the source and the surface of the sphere (which is denoted as the 'free-space condition'), the integral of the power per unit area $P(r, \theta, \phi)$ [watt/m^2] delivered in radial direction with angles θ and ϕ, must be equal to P_T

$$P_T = \iint\limits_{sphere} P(r, \theta, \phi) \, dA \qquad (3.109)$$

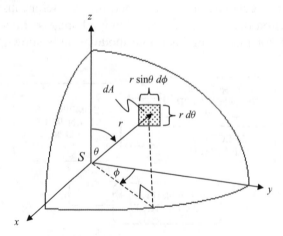

Figure 3.20 Spherical coordinates

If the source is assumed to be isotropic, i.e. it radiates in all directions with the same intensity, then $P(r, \theta, \phi)$ depends only on r. In this case, since the area of the sphere is $4\pi r^2$, (3.109) yields

$$P_T = P(r) \iint\limits_{sphere} dA = P(r) \, 4\pi r^2 \Rightarrow P(r) = \frac{P_T}{4\pi r^2} \qquad (3.110)$$

which is the well-known inverse-square law describing the power per unit area at distance r from an isotropic source.

In the general case, the radiation may be seen as generated by a number of isotropic sources with different amplitudes and phases. If the sources are concentrated near the origin, in an area of radius much smaller than r, they may be seen as a 'point source' located at the origin, radiating in radial direction with power density of the form

$$P(r, \theta, \phi) = \frac{P_T}{4\pi r^2} \, w(\theta, \phi) \; [watt/m^2] \qquad (3.111)$$

where $w(\theta, \phi)$ is some non-negative and dimensionless function describing a 'radiation pattern' independent from r. The knowledge of $w(\theta, \phi)$ and P_T supplies enough information about the nature of the source, however, before we may describe a method for their measurement, we need to define the concept of antenna aperture.

3.6.3.2 Antenna Aperture

A receiving antenna is a device capable of capturing a portion of the energy that propagates from a radiating source according to (3.111). Since the propagating energy has a power density $P(r, \theta, \phi)$ with dimensions of [watt/m²], we can say that if an antenna captures some of the radiated power, it effectively exhibits an equivalent area, named 'antenna aperture', and denoted by A [m²]. Assume that, for some fixed value of r, the antenna is positioned at (r, θ, ϕ) on a sphere with the radiating source at its center, and aligned for strongest reception. If the area A is small enough so that the power density $w(\theta, \phi)$ can be considered constant over it, then the received power $P_R(\theta, \phi)$ at the antenna output for a given set (r, θ, ϕ), can be approximated by

$$P_R(\theta, \phi) \approx A \frac{P_T}{4\pi r^2} \, w(\theta, \phi) = A \cdot P(r, \theta, \phi) \; [watt] \qquad (3.112)$$

It can be shown (Balanis, 1997) that if G is the (known) gain of the receiving antenna relative to isotropic, then

$$A = G \frac{\lambda^2}{4\pi} \; [m^2], \; \lambda = \frac{c}{f}, \; G = \max_{\theta, \phi} \{w(\theta, \phi)\} \qquad (3.113)$$

where c is the speed of light, f is the frequency of the received signal, and λ its wavelength. Substituting (3.113) into (3.112) we get

$$P_R(\theta, \phi) \approx P_T G \left(\frac{\lambda}{4\pi r}\right)^2 w(\theta, \phi) \; [watt] \tag{3.114}$$

If $P_R(\theta, \phi)$ is known for all values of θ and ϕ, then the power density $P(r, \theta, \phi)$ on the surface of the sphere is found from (3.112) and (3.113) as

$$P(r, \theta, \phi) \approx \frac{4\pi}{G\lambda^2} P_R(\theta, \phi) \; [watt/m^2] \tag{3.115}$$

and, assuming no losses, the total radiated power P_T can be approximately computed using (3.109). Noting from Figure 3.20 that

$$dA = r^2 \sin\theta \, d\theta \, d\phi, \; 0 \le \theta \le \pi, \; 0 \le \phi < 2\pi \tag{3.116}$$

by integrating $P(r, \theta, \phi)$ over a sphere of radius r, we get from (3.109)

$$P_T \approx \int_0^{2\pi} \left(\int_0^{\pi} P(r, \theta, \phi) r^2 \sin\theta \, d\theta \right) d\phi \; [watt] \tag{3.117}$$

3.6.3.3 Measurement of Radiated Spurs

By measuring the received power $P_R(\theta, \phi)$ at the antenna output for a finite set of angles $\{\theta_n, \phi_m\}$, $n = 0, 1, \ldots, N$, $m = 0, 1, \ldots M$, with the best resolution possible (ideally at least the required spatial Nyquist rate), we get a discrete set of values $\{P_R(\theta_n, \phi_m)\}$ from which the approximation to $P(r, \theta, \phi)$ is computed as

$$P(r, \theta_n, \phi_m) \approx \frac{4\pi}{G\lambda^2} P_R(\theta_n, \phi_m) \; [watt/m^2] \tag{3.118}$$

Then we may approximate the integral (3.117) in the discrete form

$$P_T \approx r^2 \Delta\theta \Delta\phi \sum_{n=0}^{N} \sum_{m=0}^{M} P(r, \theta_n, \phi_m) \sin\theta_n \; [watt]$$

$$\Delta\theta = \frac{\pi}{N}, \theta_n = n\Delta\theta, n = 0, 1, \ldots, N$$

$$\Delta\phi = \frac{2\pi}{M}, \phi_m = m\Delta\phi, m = 0, \ldots, M \tag{3.119}$$

Now, with P_T known, we may compute a discrete approximation to $w(\theta, \phi)$ from Equation (3.114)

$$w(\theta_n, \phi_m) \approx \left(\frac{4\pi r}{\lambda}\right)^2 \frac{P_R(\theta_n, \phi_m)}{P_T G} \tag{3.120}$$

In practice, the radiation pattern of most unwanted radiated signals can be considered roughly isotropic, thus we only need to find the pair of angles (θ, ϕ) for which $P_R(\theta, \phi)$ is maximal, set in (3.114) $w(\theta, \phi) = 1$, and get a reasonably tight bound to the total radiated power in the form

$$P_T \approx \left(\frac{4\pi r}{\lambda}\right)^2 \frac{1}{G} \max_{n,m} \{P_R(\theta_n, \phi_m)\} \ [watt] \qquad (3.121)$$

The measurements must be carried out at a distance of several wavelengths from the source where only the propagating field (the far-field) has substantial amplitude. Typical lab measurements for frequencies above 500 MHz are carried out at $r \approx 3$ m.

Unless a sophisticated measurement site is available, a fixed receiving antenna is used and the unit under test (UUT) is positioned on a 360° turntable. Then, the emission of the UUT is measured over a full rotation. The measurement is repeated at least three times, repositioning the UUT in mutually perpendicular directions (x, y, and z axis).

Since we do not know the polarization of the radiated field, every measurement must be taken with the receiving antenna set both for vertical and horizontal polarizations, and the measured powers are added up. Alternatively we may adjust the antenna for strongest reception and take a bound 3 dB higher than the measured value. Sometimes we are asked to provide the measurement results in terms of the average electric field in units of Volt/meter. In this case there is a simple conversion formula between the power density and the electric field (Balanis, 1997):

$$|E(r, \theta, \phi)| = \sqrt{Z_0 \cdot |P(r, \theta, \phi)|} \quad [V/m], \quad Z_0 = 377 \ \Omega \qquad (3.122)$$

where E denotes the average electric field, and Z_0 is the intrinsic impedance of free space.

Note: The very same procedure can be used for measuring the radiation pattern and the efficiency of antennas of unknown performance. In this case, P_T is the (known) power fed to the antenna, thus the approximation to $w(\theta, \phi)$ may be computed directly from (3.120). The antenna efficiency η is the ratio of the total measured radiated power to the power fed to the antenna, namely

$$\eta \approx \frac{r^2 \Delta\theta \Delta\phi}{P_T} \sum_{n=0}^{N} \sum_{m=0}^{M} P(r, \theta_n, \phi_m) \sin \theta_n \qquad (3.123)$$

3.7 Enhancement Techniques

In many applications, the 'natural' PA performance is not satisfactory. The most common issues are:

- Linearity is too poor to meet the spectral mask or the in-band noise limit dictated by the constellation density.

- Efficiency is too poor to meet the required battery operating time or the heat dissipation limits.

Several techniques have been developed to overcome the above problems. We describe here the most effective and common ones.

3.7.1 Linearization Techniques

Three main techniques are in use for PA linearization (Raab et al. (2003)

- Cartesian feedback;
- feed-forward;
- pre-distortion.

The first two are very effective and in common use. Both yield transmitter *IMD3* of the order of 60 dB and more, while the natural *IMD3* figure achievable by most 'bare' RF power amplifiers is less than 30 dB.

3.7.1.1 Cartesian Feedback

Cartesian feedback can be effectively used only for narrow channel bandwidths, usually of the order of few hundreds of kHz. There are two reasons for this limitation:

- In order to allow for proper signal processing, the bandwidth must be at most a fraction of the maximal working frequency of the baseband components.
- The implementation is based on an active feedback loop, thus we must watch out for gain and phase margins in order to avoid oscillations (Millman-Halkias, 1972). If the bandwidth is large, phase accumulation may occur either in the baseband circuits, or within RF filters and matching circuitry, eventually causing loop instability.

This technique is attractive for battery-operated portable units, such as cellular phones, since its implementation is inexpensive, adding negligible current drain and physical volume.

The feedback architecture is shown in Figure 3.21. In substance we add a receiver which detects the transmitted quadrature signals distorted by the PA, and the feedback action is carried out on the complex baseband signal rather than at RF frequency. This makes sense: feedback on two independent variables corrects both phase and amplitude simultaneously.

For the sake of simplicity, we assume that the low-pass filters in the receiving paths are ideal, and that all the signals of interest, including distortion, are within their bandwidth. If the RF bandwidth of interest is 2 Ω

$$H(\omega) = \begin{cases} 1, |\omega| < \Omega \\ 0, \text{else} \end{cases} \qquad (3.124)$$

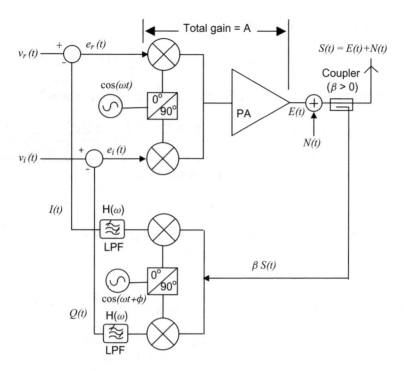

Figure 3.21   Cartesian feedback architecture

With reference to section 1.2.1.2, if the amplifier of Figure 3.21 has no distortion ($N(t) = 0$), and the feedback is disconnected ($\beta = 0$), the output signal is undistorted and has the form $S(t) = AV(t)$, where A is the total gain in the transmitter path, and

$$V(t) = Re[(v_r(t) + jv_i(t))e^{j\omega t}] \qquad (3.125)$$

If distortion occurs, it may be seen as a parasitic additive signal $N(t)$ at the output of the PA. Then the output RF signal $S(t)$ is different from $AV(t)$ and, for some $s_r(t)$ and $s_i(t)$, we may write it in the form

$$S(t) = Re[(s_r(t) + js_i(t))e^{j\omega t}] \qquad (3.126)$$

$S(t)$ is lightly sampled proportionally to $0 < \beta \ll 1$ and constitutes the incoming signal at the input of the receiver.

Assume that the phase shift ϕ has been adjusted so that the LO is exactly in-phase with the carrier of the undistorted signal, compensating for any phase shift that might accumulate in the transmit path due to filters, matching etc., then we say that the phase is 'aligned'. Phase alignment is important in ensuring loop performance and stability (Dawson and Lee, 2004). As seen in section 1.2.1.2, if the phase is aligned, the in-phase and quadrature channels are decoupled, and the recovered baseband signal has the form

$$I(t) + jQ(t) = \beta(\frac{1}{2}s_r(t) + j\frac{1}{2}s_i(t)) \qquad (3.127)$$

Then the error signal at the input of the quadrature modulator is

$$e_r(t) + je_i(t) = v_r(t) + jv_i(t) - (I(t) + jQ(t)) \qquad (3.128)$$

Using (3.125), (3.126) and (3.128), and according to Figure 3.21, the distorted output signal $S(t)$ can be expressed also in the form

$$S(t) = Re[A(e_r(t) + je_i(t))e^{j\omega t}] + N(t)$$

$$= ARe[(v_r(t) + jv_i(t))e^{j\omega t}] - \frac{\beta A}{2}Re[(s_r(t) + js_i(t))e^{j\omega t}] + N(t)$$

$$= AV(t) - \frac{\beta A}{2}S(t) + N(t) \qquad (3.129)$$

It follows from (3.129) that the distortion is divided by $1+\beta A/2$

$$S(t) = \frac{A}{(1 + \beta A/2)}V(t) + \frac{1}{(1 + \beta A/2)}N(t) \qquad (3.130)$$

Since $1/(1 + x) \approx 1 - x$ for $x \to 0$, if $\beta A \to \infty$, the distortion at output goes away

$$S(t) = \frac{2}{\beta}[V(t)] + O(\frac{2}{\beta A}) \approx \frac{2}{\beta}V(t), \beta A \gg 2 \qquad (3.131)$$

A word of caution: due to the non-ideal low-pass filters in the receiving path, we must carefully watch out for gain and phase margin, otherwise the feedback may turn positive producing loop oscillations.

3.7.1.2 Feed-forward

Technically, feed-forward can be effectively used for any channel bandwidth, however, its implementation is cumbersome, costly, current consuming, and occupies substantial physical space. Therefore it is used mostly in fixed equipment, such as wideband cellular base-station transmitters, which are less sensitive to price, current consumption and physical dimensions. As suggested by its name, feed-forward is an open-loop technique. Figure 3.22 shows an in-principle implementation of the idea before any real-life consideration is taken into account.
The process is as follows:

1. Assume that the PA has gain A. The signal at the output of the final PA is lightly sampled with attenuation equal to A, yielding a low-level version of the distorted signal with the same level as the driving signal at PA input.

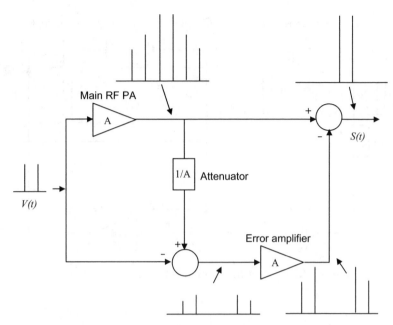

Figure 3.22 In-principle implementation of feed-forward

2. The undistorted input signal is subtracted from the sampled version. The result is the distortion signal alone, referred to as 'the error signal'. If the error signal is much smaller than the input signal, then it may be amplified by a low-power amplifier with gain A, referred to as 'the error amplifier', yielding a signal exactly equal to the distortion at the output of the PA. The amplified error signal is then subtracted from the output signal, thus canceling the distortion.

There are few obstacles in implementing this idea:

- The accuracy at which distortion can be canceled depends on the accuracy of the sampling and of the gain of the error amplifier.
- In the path going through the sampler and the error amplifier, there are time delays different from the delay experienced by the signal in the main RF path. These delays must be accurately compensated for.
- The resulting linearity is limited by the linearity of the components participating in the process, among them the error amplifier.
- A calibration process must be continuously carried out and adjusted to compensate for voltage and temperature drift, and other variations that occur both in the PA and in the correcting circuits.

The result is a rather complicated machine, driven by complex algorithms. Figure 3.23 describes a real-life implementation. In spite of its complexity, feed-forward is the best choice whenever wideband operation is required, and its open-loop character guarantees no danger of oscillation.

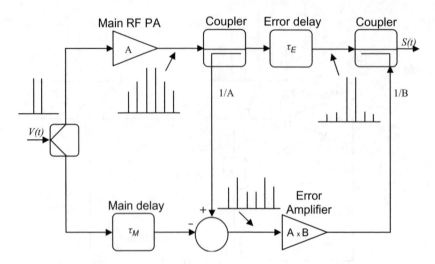

Figure 3.23   Real-life implementation of feed-forward

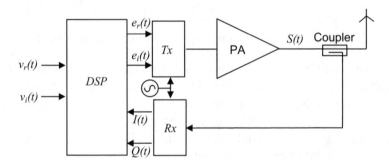

Figure 3.24   Real-life implementation of pre-distortion

3.7.1.3 Pre-distortion

Pre-distortion is less commonly used, and is heavily based on computation-intensive signal processing. Its basic topology, shown in Figure 3.24, is somewhat similar to the Cartesian feedback (compare Figure 3.21).

The idea is the following: the quadrature output from the sampling receiver is analyzed and compared with the undistorted input. Then we compute an 'inverse' nonlinear forward DSP transfer function looking for the minimal error power at output, or some other optimization criteria.

There are many possible algorithmic approaches to pre-distortion, and their treatment is beyond the scope of this book.

3.7.2 Envelope-Tracking Supply

Most power amplifiers traditionally operate under constant supply voltage, and reach maximum efficiency in the vicinity of power saturation.

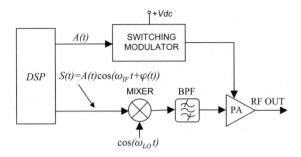

Figure 3.25   Envelope-tracking supply

We saw before that when the transmitted signal has high peak to average ratio, the average power is far less than the saturation value, which usually results in substantial efficiency degradation. From the analysis of section 3.4, it follows that we could dramatically improve PA efficiency by making the DC supply voltage follow the instantaneous amplitude of the envelope of the transmitted RF signal. Indeed, this would constantly keep the PA near its saturation point, which corresponds to the best efficiency condition. Figure 3.25 describes the basic concept of envelope-tracking (ET) supply.

The DSP provides two outputs: the (properly delayed) baseband envelope $A(t)$ and the amplitude and phase-modulated signal at IF frequency. The baseband envelope in then routed to the ET modulator. The ET modulator is in essence a dynamic UP/DOWN switching converter, which tracks and amplifies the input envelope with low distortion, low delay time, and high efficiency. The modulated signal at IF is mixed up to RF frequency by the local oscillator, bandpass-filtered to eliminate the image signal and fed to the final RF PA. The PA can have any type of linear architecture, including feedback or not.

4

Synthesizers

Modern digital transceivers for data communication applications use advanced air interfaces requiring both frequency agility and good spectral purity. One outcome is that synthesizer performance is becoming more and more critical. The demand for small size, fast switching time, along with a 'clean' spectrum and low power consumption leads to sophisticated synthesizer architectures.

In the following, we assume that the reader is familiar with basic synthesizer theory. Nevertheless, for the sake of completeness, we include later in this chapter a detailed review of synthesizer fundamentals, and we show how to extend their validity to advanced 'pseudorandom-fractional' synthesizers.

4.1 Synthesizer Architectures

In modern equipment, frequency synthesizers are used whenever a system requires the use of a time clock, for example in communications and signal processing applications. In RF transceivers, the local oscillators (LO) are implemented using a digital synthesizer. As seen in Chapter 2, the LO is a major building block that critically affects the total transceiver performance.

A general insight and overview on synthesizers can be found in Rohde (1997). Oscillator/VCO theory is extensively treated in Chapter 5.

The most common synthesizer architecture, shown in Figure 4.1, comprises

- A free-running voltage-controlled oscillator (VCO).
- A fixed 'reference' oscillator.
- A dynamically programmable digital counter.
- A phase/frequency detector.
- An analog low-frequency feedback network, referred to as 'the loop filter'.

The free-running VCO is a device capable of oscillating over the required RF frequency range, and whose frequency can be controlled through a port often referred to as 'the steering line'. The reference oscillator is a frequency-precise,

Wireless Transceiver Design Ariel Luzzatto and Gadi Shirazi
© 2007 John Wiley & Sons, Ltd

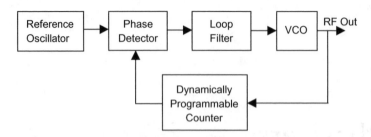

Figure 4.1   In-principle architecture of a digital synthesizer

low-noise, temperature-stable and low-aging device, mostly based on piezoelectric crystal technology. Usually, in order to attain the required stability and noise characteristics, it oscillates at a frequency much lower than the VCO.

The digital portion of the synthesizer includes a high-speed dynamically programmable counter, which divides the VCO frequency down to a value very close to the reference frequency. The counter is not necessarily fixed, but may divide the RF frequency by a variable integer whose value may change 'on the fly', following some predefined pattern.

The phase detector is a digital device that compares, at the reference rate, the instantaneous 'rising edge' time (phase) difference between the reference oscillator signal and the frequency-divided VCO signal, and outputs an 'error' pulse whose width and sign is equal to this time difference. Thus the error signal consists of a train of (usually narrow) pulses at a rate equal to the reference frequency. If the error signal exceeds certain limit conditions, the phase detector enters an operating mode referred to as the 'frequency detection mode'.

The loop filter is a low-pass filter that averages the error signal from the phase detector, and feeds the steering line of the VCO with a correcting signal depending on the short-term average phase error.

Using the above arrangement, the free-running VCO is able to oscillate at many different frequencies in its range, while exhibiting the stability and purity characteristics of the reference oscillator. When this happens, we say that the VCO is 'locked' to the reference oscillator. Since, as we see later, VCO locking is usually achieved by making the phase of the divided RF signal follow, on the average, the phase of the reference oscillator, a synthesizer of the type shown in Figure 4.1 is also referred to as a 'phase locked loop' (PLL).

Since the counter is programmable to different division ratios, but the divided VCO signal is always compared to the same fixed reference oscillator frequency, the direct consequence is that VCO locking occurs at a frequency dependent on the count number.

Most modern frequency synthesizers utilize PLL techniques. The required PLL characteristics, however, are strongly dependent on the specific application. For instance, a synthesizer for a Bluetooth (BT) transceiver, which must operate in fast frequency hopping (FH) mode (selecting a different carrier frequency at the rate of 1600 'frequency hops' every second), will be very different from one that needs to operate in a code division multiple access (CDMA) transceiver, which needs to change frequency only sporadically. The required spectral purity is also dependent on the air protocol used.

Needless to say, speed, spectral purity, size, complexity, current consumption and cost are all conflicting requirements, thus, PLL performance must be tailored to each specific application.

It is customary to classify the different types of synthesizers according to the way the programmable divider works. Until 15 years ago, the only PLL architecture found in the market was the 'integer-N PLL', where the counter divides the VCO frequency by a fixed integer number. This is the simplest architecture but has several severe limitations and is being abandoned.

A more sophisticated approach is the 'fractional-N PLL' where the counter divides the VCO frequency by an integer number that varies over time following some given pattern. The result is that the VCO frequency is effectively divided by a rational (fractional) number. The integer-N PLL can be seen, in principle, as a particular case of fractional-N (although the practical implementation is different). The fractional-N has many advantages but suffers from spectral impurity.

Improved implementations of the fractional-N approach, such as MASH (Multi stAge noise SHaping) architectures, generate division patterns according to pseudorandom sequences, and exhibit good spectral purity. The 'MASH Sigma-Delta fractional-N PLL' architecture is the most useful and common, and we treat it extensively here.

An all-digital, non-PLL (and complex) architecture named DDS (Direct Digital Synthesis) did not find many applications in industrial and commercial transceiver design, and will not be treated here.

Before proceeding further, we suggest reading section 4.7, in order to refresh basic PLL theory.

4.2 Fractional-N Outlook

In section 4.1, we pointed out that the phase detector generates pulses at a rate equal to the reference oscillator frequency. We also pointed out that the loop filter is a low-pass filter whose purpose, among others, is to average the train of error pulses from the detector. We recall the following from PLL fundamentals:

- A WIDE LOOP FILTER IS GOOD: the synthesizer lock time (the time required to 'jump' from one frequency to another within a predefined error) is closely related to the inverse of the bandwidth of the loop filter. The wider the loop-filter bandwidth, the shorter the lock time can be made.
- A WIDE LOOP FILTER IS BAD: the loop-filter bandwidth is too wide, the harmonic components of the pulses generated by the phase detector will reach the steering line, and will frequency-modulate the VCO generating a spectral

interference known as 'reference spur' which causes, among other things, selectivity and blocking deterioration in the receiver.

• A GOOD LOOP FILTER IS BAD: the loop filter cannot be sharp, because if it contains too many poles, it will jeopardize synthesizer stability ending up with parasitic PLL oscillations. Thus the filtering action of the loop filter is generally poor.

It follows from the above, that lock-up speed, spectral purity and stability are conflicting requirements. As we see later, the best way to minimize the conflict is to reduce as much as possible the dividing value of the programmable counter, namely use a high reference frequency. However, if the programmable counter is set to divide by a fixed value, as is the integer-N approach, the basic PLL theory shows that the VCO may lock only on a multiple of the reference frequency.

THE BAD NEWS

• A small divider value, implying a high reference frequency, yields a poor frequency resolution (the multiples of the reference frequency are spaced far away from each other). Thus integer-N synthesizers cannot provide fast lock-up times and fine frequency resolution at the same time.

THE GOOD NEWS

• The fractional-N approach solves the above contradictory requirements by using a programmable counter that changes the divide value 'on-the-fly' according to a given pattern. The idea beyond the fractional-N strategy is:
  – Increase the reference frequency far beyond the 3 dB corner of the loop filter, so that even with a relatively large filter bandwidth and with moderate filter sharpness, the reference spurs will be filtered out.
  – Since the VCO frequency is effectively divided by a fractional number, the VCO may now lock on to frequencies that are multiples of *a fraction* of the reference. Thus, in spite of the high reference frequency, the synthesizer resolution will be good, along with a fast lock-up time and while retaining spectral purity.

As a result of the variable divide value, the *divided* VCO signal becomes phase-modulated, but, as we will see shortly, under lock condition the *average* divided frequency will be the reference frequency, and the ratio between VCO and reference frequencies will be a rational number.

4.3 Fractional-N Theory

Assume that the programmable counter of Figure 4.1 sets its nth divide value to N_n, according to a periodic counting pattern of integer period D.

Let $f_0 = 1/T_0$ be the VCO frequency, and $f_R = 1/T_R$ be the reference oscillator frequency, and assume that during the full divide pattern, the cumulative lead or lag time of the rising edge of the *divided* VCO signal with respect to the rising edge of the reference oscillator signal, never exceeds half the reference period (this will be justified later on).

Under this assumption, the rising edge of one signal appears exactly once during each period of the other signal. Thus, during the pattern period, the phase detector produces exactly D pulses, whose width and polarity match the lead/lag times between the rising edges of the two signals. The polarity of the pulses is (arbitrarily) assumed to be positive if the divided VCO edge leads in time, and negative otherwise.

The cumulative time error E_D over a full counting pattern period, relative to the reference oscillator, is therefore

$$E_D = \sum_{n=1}^{D}(N_n T_0 - T_R) = \sum_{n=1}^{D} N_n T_0 - D T_R, \; T_0 = \frac{1}{f_0}, \; T_R = \frac{1}{f_R} \quad (4.1)$$

Usually, the phase-detector/loop-filter architecture is such that lock-up occurs when the cumulative time error over D pulses is zero. Thus, under lock-up conditions

$$E_D = \sum_{n=1}^{D} N_n T_0 - D T_R = 0 \quad (4.2)$$

which yields

$$\frac{f_0}{f_R} = \frac{T_R}{T_0} = \frac{1}{D}\sum_{n=1}^{D} N_n \quad (4.3)$$

Since $\{N_n\}$ and D in Equation (4.3) are integers, then f_0/f_R is a rational (fractional) number. If we set

$$N_n = N + \Delta_n, \; |\Delta_n| \in \{0, 1, 2, \ldots\}, \quad M = \sum_{n=1}^{D} \Delta_n < D \quad (4.4)$$

for the lock-up condition $E_D = 0$, Equation (4.2) becomes

$$E_D = \left(DN + \sum_{n=1}^{D} \Delta_n\right) T_0 - D T_R = 0 \quad (4.5)$$

and Equation (4.3) yields

$$\frac{f_0}{f_R} = N + \frac{M}{D}, \; \frac{M}{D} < 1 \quad (4.6)$$

4.3.1 Dual-Count Fractional-N

The simplest fractional-N implementation uses the count pattern

$$\Delta_{q_1} = \Delta_{q_2} = \cdots = \Delta_{q_M} = 1$$

$$\Delta_{q_{M+1}} = \Delta_{q_{M+2}} = \cdots = \Delta_{q_D} = 0$$

$$\{q_j\}_{j=1}^D \in \{1, 2, \ldots, D\}, D, M, N \in \{1, 2, \ldots\} \tag{4.7}$$

Note that the indexing $\{q_j\}$ implies that the count is set M times at the value $N + 1$ and $D - M$ times at the value N, but the switching between the two values may appear in any order. Using (4.7) we get

$$\sum_{n=1}^D \Delta_n = M \tag{4.8}$$

Thus it follows from (4.6) that

$$f_0 = \left(N + \frac{M}{D}\right) f_R, \quad \frac{M}{D} < 1 \tag{4.9}$$

4.3.2 First-Order Sigma-Delta Fractional-N

As pointed out in the previous section, the fractional division ratio of the dual-count synthesizer holds regardless of the switching pattern, provided that the count value is set M times to $N + 1$, and $D - M$ times to N. However, the harmonic content of the error signal delivered to the VCO (and consequently the spectral purity of the synthesizer output) is very much dependent on the way the counting sequence is implemented.

Since the loop filter is of low-pass type, good filtering calls for counting patterns such that the energy of the spurious content of the error signal is concentrated as high as possible in frequency. In other words, we aim for a rapidly varying counting pattern. Such a distribution of spurious energy, named 'noise shaping', is accomplished by controlling the divide value of the counter using pseudorandom sequences.

In the dual-count case, the values of $\{\Delta_n\}$, $n = 1, \ldots, D$ in (4.7) are set to be equal to a binary pseudorandom sequence. A convenient way of generating such a binary sequence is the use of a first-order digital sigma-delta modulator.

The sigma-delta modulator accepts a constant integer input value M, $0 \le M < D$ and a clock, and outputs a sequence of carry words $\{C_n\}$ generated by the summing operation and defined in (4.10), from which the binary sequence $\{\Delta_n\}$ is derived.

With an arbitrary integer initial value $0 \le Q_0 < D$, and after n clocks, the modulator transfer function is

$$Q_n = (M + Q_{n-1}) \bmod D = M + Q_{n-1} - C_n \tag{4.10}$$

where

$$C_n = 0 \text{ if } M + Q_{n-1} < D$$

$$C_n = D \text{ if } M + Q_{n-1} \geq D$$

$$0 \leq \{M, Q_n\} < D, \Delta_n = C_n/D$$

From the equations above, it is evident that if $M + Q_{n-1} \geq D$ then $Q_n < Q_{n-1}$ and if $M + Q_{n-1} < D$ then $Q_n > Q_{n-1}$, thus the modulator tends to generate a rapidly varying sequence.

At this point, recall the following from modular arithmetic (Rozen, 2003):

- **A mod N**: let A and N be positive integers. Then $A \bmod N$ is the residual of the division A/N.

$$A \bmod N = res(A/N) \tag{4.11}$$

For instance $17/5 = 3 + 2/5 \Rightarrow 17 \bmod 5 = 2$.

- **Commutativity** and **associativity** hold, under modular arithmetic, for both addition and multiplication. **Distributivity** holds as well.
- **Reducibility**:

$$(A + B) \bmod N = [(A \bmod N) + (B \bmod N)] \bmod N$$

$$(A \times B) \bmod N = [(A \bmod N) \times (B \bmod N)] \bmod N \tag{4.12}$$

For the sake of simplicity, let us use from now on the notation

$$[A]_B \equiv A \bmod B \tag{4.13}$$

Therefore (4.10) becomes

$$Q_n = [M + Q_{n-1}]_D \tag{4.14}$$

Thus, starting with Q_0, since $Q_0 = [Q_0]_D$ and $M = [M]_D$, in view of the reducibility property (4.12), we get

$$Q_1 = [[M]_D + [Q_0]_D]_D = [M + Q_0]_D$$

$$Q_2 = [[M]_D + [Q_1]_D]_D = [[M]_D + [M + Q_0]_D]_D = [2M + Q_0]_D$$

$$\vdots$$

$$Q_n = [nM + Q_0]_D \tag{4.15}$$

Again by the reducibility property, the last equation in (41.5) yields

$$Q_n = [nM + Q_0]_D = [[nM]_D + [Q_0]_D]_D = [[nM]_D + Q_0]_D \tag{4.16}$$

If there exists some integer $k > 0$ for which $Q_{n+k} = Q_n$ for *all* values of n, then the sequence $\{Q_n\}$ is periodic with period k. Equation (4.16) implies that

$$Q_{n+k} = [nM + kM + Q_0]_D = [nM + Q_0]_D \Rightarrow [kM]_D = 0 \qquad (4.17)$$

which in turn means that kM must be an integer multiple of D.

Clearly if $k = D$ Equation (4.17) holds, thus the period of the binary sequence $\{\Delta_n\}$ is *at most* D. However, there exist sequences with a period shorter than D. To illustrate this, denote by $gcd(M, D) \geq 1$ the greatest common divider of M and D. Then there exists some integer number M' such that

$$M' = \frac{M}{gcd(M, D)} \qquad (4.18)$$

If $k = D'$ and D' satisfies (4.19)

$$D' = \frac{D}{gcd(M, D)} \qquad (4.19)$$

we get

$$[kM]_D = [D'M]_D = [M'D]_D = 0 \qquad (4.20)$$

Equation (4.20) shows that adding D' times the input value M yields the same result as if we added M' times a value D. Under modulo-D summation, and since $M < D$, then $M'D$ generates a non-zero carry M' times. Thus the binary sequence $\{\Delta_n\}$ has M' ones and $D' - M'$ zeros. By (4.16)

$$Q_{D'} = [D'M + Q_0]_D = [M'D + Q_0]_D = [[M'D]_D + Q_0]_D = Q_0 \qquad (4.21)$$

Therefore the period of the sequence $\{\Delta_n\}$ is $D' \leq D$. If M and D are not relatively prime, then $gcd(M, D) > 1$ and $D' < D$.

Summarizing: starting from the value Q_0, a first-order sigma-delta modulator should perform the modulo-D addition of the value M to the previous result and output the carry bit (note from (4.10) that the value $\Delta_n = C_n/D$ is just the carry bit of the nth addition).

Figure 4.2 shows a (non-unique) implementation of a first-order sigma-delta fractional-N counter.

- The adder performs modulo-D addition.
- The output of the counter $\{P_n\}$ acts as a clock triggering the adder action.
- The one-bit carry Δ_n is either 0 or 1, thus the count is N or $N + 1$ accordingly.

As pointed out before, the counting pattern is periodic, thus, if D, M and N are given, the output waveform $\{P_n\}$ is periodic too.

In the following we denote by a 'clock', the waveform included between two consecutive rising edges of $\{P_n\}$. We should be aware, however, that since the

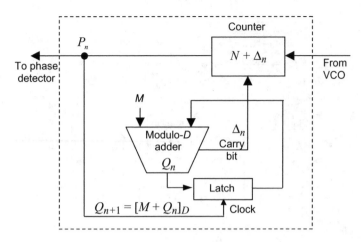

Figure 4.2 First-order sigma-delta implementation

count value varies according to the value of Δ_n, different clocks may have different time lengths.

The adder overflows M' times, thus the counter has been set to $N + 1$ during M' output clocks, and to N during $D' - M'$ output clocks in the switching sequence generated by the sigma-delta modulator.

We summarize the findings:

- The period of the counting pattern is $D' \leq D$ clocks, with D' given in (4.19).
- If M and D are not relatively prime $D' < D$, else $D' = D$.
- The counter is set to $N + 1$ during M' clocks, and to N during $D' - M'$ clocks.

Getting back to the results of section 4.3, we may now conclude from (4.9), (4.18) and (4.19), that using the counter of Figure 4.2 with the synthesizer of Figure 4.1, we get

$$\frac{f_0}{f_R} = N + \frac{M'}{D'} = N + \frac{M}{D} \tag{4.22}$$

4.4 Multi Stage Noise Shaping (MASH) Architecture

MASH architectures are of great importance, and in spite of their complexity they deserve the instructive in-depth analysis that follows.

A third-order MASH architecture using modulo-D adders is shown in Figure 4.3. Higher-order MASH design is easily inferred from it.

The pseudorandom sequence generated by the simple sigma-delta modulator described in the previous section does not produce good noise spreading for all values of $0 < M < D$. For instance, it is easy to verify that, for $M = 1$, the carry bit is set only once every D counts, resulting in strong low-frequency spurious content on the steering line.

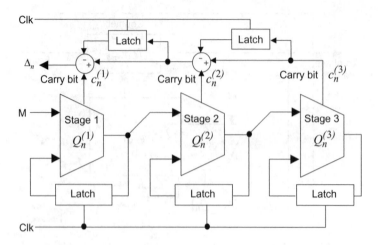

Figure 4.3   Third-order MASH architecture using modulo-D adders

It is possible to generate 'better' pseudorandom sequences using several first-order sigma-delta modulators in the 'nested' architecture of Figure 4.3 called MASH (Multi stAge noise SHaping). The number N of the nested modulators is called the 'order' of the pseudorandom sequence generator, and the programmable divider is now controlled by a multi-valued sequence $\{\Delta_n\}$.

In order to illustrate the operation of the MASH architecture we analyze in detail the behavior of the third-order modulator above. However, it is straightforward to extend the following analysis to any higher MASH order.

4.4.1 Stage One

The first stage is the first-order digital sigma-delta modulator analyzed in section 4.3.2. In all practical implementations D is a power of two. *Thus, if we chose M to be odd, D and M are relatively prime.*

The consequence is that $\gcd(M, D) = 1$, and the period of the counting pattern of stage one is of maximal length D.

$$D = 2^p, \quad M = 2q - 1, \quad p, q \in 1, 2, 3, \ldots \Rightarrow gdc(M, D) = 1 \qquad (4.23)$$

If we double the value of D, the choice of odd values of M does not impair the resolution of the synthesizer. Thus, from now on, we may always assume $D' = D$. Then, as pointed out in the previous section, the adder overflows M times, namely, the carry bit is set M times every D clocks.

Taking arbitrarily $Q_0^{(1)} = 0$ (which has no effect except introducing a time delay) and using (4.16), we get

$$Q_n^{(1)} = [nM]_D, \quad n = 0, 1, \ldots, D - 1 \qquad (4.24)$$

Since M and D are relatively prime, their least common multiple (lcm) is

$$lcm(M, D) = DM \qquad (4.25)$$

Therefore, for $p = 0, 1, \ldots, D - 1$, pM cannot be a multiple of D. By (4.24)

$$Q_{n+p}^{(1)} = [(n + p)M]_D = [Q_n^{(1)} + [pM]_D]_D \neq Q_n^{(1)}$$
$$n + p = 0, 1, \ldots, D - 1 \tag{4.26}$$

The consequence of (4.26) is that, since the output of the adder can generate only D different values, $\{Q_n^{(1)}\}$ spans, in *some* order, *all* the values $0, 1, \ldots, D - 1$, and each value appears *only once* during D subsequent clocks. We may summarize

$$Q_n^{(1)} \in \{0, 1, \cdots, D - 1\}, \quad n \in \{0, 1, \cdots, D - 1\}$$
$$Q_{n+p}^{(1)} \neq Q_n^{(1)}, Q_{n+D}^{(1)} = Q_n^{(1)}, n + p = 0, 1, \ldots, D - 1$$

carry bit set M times over D clocks $\tag{4.27}$

4.4.2 Stage Two

Unlike stage one, the input of stage two is not constant, but consists of the sequence $\{Q_n^{(1)}\}$. Therefore, the output $\{Q_n^{(2)}\}$ of stage two consists of the partial sums of the output values of stage one plus a 'seed' (starting) value $Q_s^{(2)}$, namely

$$Q_q^{(2)} = \left[\sum_{m=0}^{q-1} Q_m^{(1)} + Q_s^{(2)} \right]_D, \quad q \in \{0, 1, 2, \ldots\}, \ b < a \Rightarrow \sum_{m=a}^{b} (\cdot) \equiv 0 \tag{4.28}$$

We note that the summing convention in (4.28) implies that $Q_0^{(2)} = Q_s^{(2)}$. Setting $q = n + kD, 0 \leq n < D, k \in \{0, 1, 2 \ldots\}$, we may rewrite (4.28) as

$$Q_{n+kD}^{(2)} = \left[\sum_{m=0}^{n+kD-1} Q_m^{(1)} + Q_s^{(2)} \right]_D$$
$$= \left[\sum_{p=1}^{k} \sum_{m=(p-1)D}^{pD-1} Q_m^{(1)} + \sum_{m=kD}^{kD+n-1} Q_m^{(1)} + Q_s^{(2)} \right]_D \tag{4.29}$$

Recalling from (4.21) and (4.27) that the period of $\{Q_n^{(1)}\}$ is D, during which all the values $0, 1, \ldots, D - 1$ appear once, then

$$\sum_{m=(p-1)D}^{pD-1} Q_m^{(1)} = \sum_{m=0}^{D-1} Q_{m+(p-1)D}^{(1)} = \sum_{m=0}^{D-1} Q_m^{(1)} = \sum_{m=0}^{D-1} m = \frac{(D-1)D}{2}$$

$$\sum_{m=kD}^{kD+n-1} Q_m^{(1)} = \sum_{m=0}^{n-1} Q_{m+kD}^{(1)} = \sum_{m=0}^{n-1} Q_m^{(1)}, \ n < D, \ k \in \{0, 1, 2 \ldots\} \tag{4.30}$$

Using (4.30) and (4.28), Equation (4.29) takes the form

$$Q_{n+kD}^{(2)} = \left[k\frac{(D-1)D}{2} + \sum_{m=0}^{n-1} Q_m^{(1)} + Q_s^{(2)} \right]_D = \left[\left[k\frac{(D-1)D}{2} \right]_D + Q_n^{(2)} \right]_D$$
(4.31)

It is easy to see from (4.31) that the smallest integer satisfying $Q_{n+kD}^{(2)} = Q_n^{(2)}$ for *all* values of n is $k = 2$. Indeed $[(D-1)D]_D = 0$, and (4.31) implies that $\{Q_n^{(2)}\}$ is periodic with period $2D$.

$$Q_{n+2D}^{(2)} = Q_n^{(2)}, \quad n = 0, 1, \ldots, D-1.$$
(4.32)

With $k = 2$, (4.31) shows that the adder overflows $D-1$ times, thus the carry bit is set $D-1$ times every $2D$ clocks, independently from the value of M.

However, there is more about $Q_n^{(2)}$. From (4.28) we observe that

$$Q_{p+q}^{(2)} = \left[\sum_{m=0}^{p+q-1} Q_m^{(1)} + Q_s^{(2)} \right]_D = \left[\sum_{m=p}^{p+q-1} Q_m^{(1)} + \sum_{m=0}^{p-1} Q_m^{(1)} + Q_s^{(2)} \right]_D$$

$$= \left[\left[\sum_{m=p}^{p+q-1} Q_m^{(1)} \right]_D + Q_p^{(2)} \right]_D$$
(4.33)

In view of (4.24), (4.25) and the analysis of stage one, if $1 \le q \le D$ then

$$\sum_{r=0}^{q-1} Q_r^{(1)} = M\frac{(q-1)q}{2} \ne kD, \quad k = 1, 2, 3, \ldots.$$
(4.34)

since either $M(q-1)$ or Mq is odd, and

$$\left[\sum_{r=0}^{q-1} Q_r^{(1)} \right]_D \ne 0, \quad 1 < q \le D$$
(4.35)

since (4.35) fails to hold only if $q = 1 \Rightarrow r \equiv 0$ and the sum consists of one element of the form $Q_{kD}^{(1)} = Q_0^{(1)} = 0$.

Similar considerations show that, if $1 \le p+q \le D$ and $q \ge 1$, the sum in the last line of (4.33) is neither zero nor a multiple of D. Indeed

$$2p + q - 1 \le p + D - 1 < 2D \Rightarrow \sum_{m=p}^{p+q-1} Q_m^{(1)} = Mq\frac{(2p+q-1)}{2} \ne kD$$
(4.36)

If we set

$$p + q = 1, 2, \ldots, D-1, \quad q \ge 1$$
(4.37)

Equation (4.36) implies that the elements $\{Q_n^{(2)}\}, n = 0, 1, 2, \ldots, D-1$ of the sequence are all different.

Since the output of the adder can generate only D different values, then $\{Q_n^{(2)}\}$ spans, in *some* order, *all* the values $0, 1, \ldots, D-1$, and each value appears *only once* during the above D subsequent clocks.

The very same reasoning may be used setting

$$p + q = D + 1, D + 2, \ldots, 2D - 1, \quad q \geq 1 \tag{4.38}$$

and it follows that also the remaining set of D elements, namely $\{Q_n^{(2)}\}$, $n = D, D+1, \ldots, 2D-1$, are all different and span *all* the values $0, 1, \ldots, D-1$ in *some* order.

We conclude that the sequence $\{Q_n^{(2)}\}$, $n = 0, 1, 2, \ldots, 2D-1$ contains each of the values $0, 1, \ldots, D-1$ *exactly two times* over $2D$ consecutive clocks.

There is one more interesting property: setting in (4.33)

$$p = D - n, \quad q = 2n + 1$$

we may write

$$Q_{D+n+1}^{(2)} = \left[\sum_{r=0}^{2n} Q_{r+D-n}^{(1)} + Q_{D-n}^{(2)} \right]_D = \left[\left[\sum_{r=0}^{2n} Q_{r+D-n}^{(1)} \right]_D + Q_{D-n}^{(2)} \right]_D \tag{4.39}$$

Now, substituting into (4.39)

$$m = r + D - n$$

and with the help of (4.24) we note that

$$\left[\sum_{r=0}^{2n} Q_{r+D-n}^{(1)} \right]_D = \left[\sum_{m=D-n}^{D+n} Q_m^{(1)} \right]_D = \left[\sum_{m=D-n}^{D-1} Q_m^{(1)} + \sum_{m=D}^{D+n} Q_m^{(1)} \right]_D$$

$$= \left[\sum_{m=D-n}^{D-1} [mM]_D + \sum_{m=0}^{n} [mM]_D \right]_D = \left[M \sum_{m=D-n}^{D-1} m + M \sum_{m=0}^{n} m \right]_D$$

$$= \left[M \frac{(D-n) + (D-1)}{2} n + M \frac{(n+1)}{2} n \right]_D$$

$$= \left[M \frac{2D - (n+1)}{2} n + M \frac{(n+1)}{2} n \right]_D = [nMD]_D = 0 \tag{4.40}$$

and therefore (4.39) yields

$$Q_{D+n+1}^{(2)} = Q_{D-n}^{(2)}, \quad n = 0, 1, \ldots, D-1 \tag{4.41}$$

The properties of $\{Q_n^{(2)}\}$ are summarized below

$$Q_n^{(2)} \in \{0, 1, \ldots, D-1\}, \quad n = 0, 1, \ldots, 2D-1$$

$$Q_{n+2D}^{(2)} = Q_n^{(2)}, \quad n = 0, 1, \ldots, 2D-1$$

$$Q_n^{(2)} \neq Q_m^{(2)}, \ n \neq m, \ n, m \in \{1, \ldots, D\}$$

$$Q_n^{(2)} \neq Q_m^{(2)}, \ n \neq m, \ n, m \in \{0, D+1, \ldots, 2D-1\}$$

$$Q_{D+n+1}^{(2)} = Q_{D-n}^{(2)}, \ n = 0, 1, \ldots, D-1$$

carry bit set $D-1$ times over $2D$ clocks   (4.42)

4.4.3 Stage Three

The input of stage three consists of the sequence $\{Q_n^{(2)}\}$. Therefore, the output $\{Q_n^{(3)}\}$ of stage three consists of the partial sums of the output values of stage two plus a seed value $Q_s^{(3)}$, namely

$$Q_q^{(3)} = \left[\sum_{m=0}^{q-1} Q_m^{(2)} + Q_s^{(3)} \right]_D, \ q \in \{0, 1, 2, \ldots\}, \ b < a \Rightarrow \sum_{m=a}^{b} (\cdot) \equiv 0 \quad (4.43)$$

As in stage two, the summing convention in (4.43) implies that $Q_0^{(3)} = Q_s^{(3)}$. Since $\{Q_n^{(2)}\}$ is periodic with period $2D$, the period of $\{Q_n^{(3)}\}$ is at least $2D$. Indeed $\{Q_n^{(3)}\}$ is of period $2D$ as well.

We show that using (4.43) and setting

$$q = n + kD, \ 0 \leq n < 2D, \ k \in \{0, 1, 2 \ldots\}$$

we get

$$Q_{n+k2D}^{(3)} = \left[\sum_{m=0}^{n+k2D-1} Q_m^{(2)} + Q_s^{(3)} \right]_D$$

$$= \left[\sum_{p=1}^{k} \sum_{m=(p-1)2D}^{p2D-1} Q_m^{(2)} + \sum_{m=k2D}^{k2D+n-1} Q_m^{(2)} + Q_s^{(3)} \right]_D \quad (4.44)$$

Recalling from (4.42) that the period of $\{Q_n^{(2)}\}$ is $2D$, during which all the values $0, 1, \ldots, D-1$ (the values of $\{Q_n^{(1)}\}$) appear twice, then

$$\sum_{m=(p-1)2D}^{p2D-1} Q_m^{(2)} = \sum_{m=0}^{2D-1} Q_{m+(p-1)2D}^{(2)} = 2 \sum_{m=0}^{D-1} Q_m^{(1)} = 2 \sum_{m=0}^{D-1} m = (D-1)D$$

$$\sum_{m=k2D}^{k2D+n-1} Q_m^{(2)} = \sum_{m=0}^{n-1} Q_{m+k2D}^{(2)} = \sum_{m=0}^{n-1} Q_m^{(2)}, \ n < 2D, \ k \in \{0, 1, 2 \cdots\} \quad (4.45)$$

With the help of (4.45), Equation (4.44) becomes

$$Q_{n+k2D}^{(3)} = \left[k(D-1)D + \sum_{m=0}^{n-1} Q_m^{(2)} + Q_s^{(3)} \right]_D = Q_n^{(3)}, \quad k \geq 1 \qquad (4.46)$$

Since $k = 1$ satisfies (4.46) for all n, then $\{Q_n^{(3)}\}$ is of period $2D$. Equation (4.46) also shows that, as in the case of stage two, the adder overflows $D-1$ times, and thus the carry bit is set $D-1$ times within $2D$ clocks, independently from the value of M. Here we have no need to find the sequence $\{Q_n^{(3)}\}$ itself, but we will have to compute some statistical properties about it in the next section. In particular we show in (4.68) that $\{Q_n^{(3)}\}$ are uniformly distributed on $[0, D-1]$. More generally, we will show that if the input to a modulo-D adder is uniformly distributed on $[0, D-1]$, so is its output, which allows straightforward MASH order extension. For now, the relevant property is

$$Q_{n+2D}^{(3)} = Q_n^{(3)}, \quad n = 0, 1, \ldots, 2D-1$$

4.5 MASH Noise Analysis

The interconnections described in Figure 4.3 have the purpose of manipulating the carry bits so as to increase the variations of the counter setting, thus pushing the noise of the signal delivered to the VCO steering line towards high frequencies. To understand the need for this noise-shaping arrangement, recall that in a first-order sigma-delta synthesizer, the carry bit is set M times every D clocks. If M is set to 1, then the carry bit of Figure 4.2 is 0 most of the time thus generating a low-frequency spur on the steering line. Since the loop filter of the PLL is low-pass, strong spurs are generated depending on the value of M. The MASH architecture redistributes the spectral content of the steering voltage spurs towards the range where the loop filter has high attenuation. Denoting by d the delay operator, namely $d(x_n) = x_{n-1}$, the value of the carry bit combination in Figure 4.3 is

$$\Delta_n = (1-d)^2 c_n^{(3)} + (1-d)c_n^{(2)} + c_n^{(1)} \qquad (4.47)$$

Now, as done in (4.10) of section 4.3.2, we may write for all n

$$Q_n^{(k)} = [Q_n^{(k-1)} + Q_{n-1}^{(k)}]_D = Q_n^{(k-1)} + Q_{n-1}^{(k)} - C_n^{(k)}$$
$$C_n^{(k)} = c_n^{(k)} D \in \{0, D\}, \, Q_n^{(0)} = M, k = 0, 1, 2, 3 \qquad (4.48)$$

from which

$$C_n^{(k)} = Q_n^{(k-1)} + (d-1)Q_n^{(k)} \qquad (4.49)$$

Using (4.49), it is easy to show by induction that

$$\sum_{k=1}^{K} (1-d)^{k-1} C_n^{(k)} = Q_n^{(0)} - (1-d)^K Q_n^{(K)} \qquad (4.50)$$

indeed, for $K = 1 \Rightarrow C_n^{(1)} = Q_n^{(0)} + (d-1)Q_n^{(1)}$, which is just (4.49). Then assuming that (4.50) holds for K, and using this induction assumption, we get

$$\sum_{k=1}^{K+1} (1-d)^{k-1} C_n^{(k)} = Q_n^{(0)} - (1-d)^K Q_n^{(K)} + (1-d)^K C_n^{(K+1)}$$

$$= Q_n^{(0)} - (1-d)^K Q_n^{(K)} + (1-d)^K [Q_n^{(K)} + (d-1)Q_n^{(K+1)}]$$

$$= Q_n^{(0)} - (1-d)^{K+1} Q_n^{(K+1)} \tag{4.51}$$

Therefore, using (4.50) with $c_n^{(k)} = C_n^{(k)}/D$ and $Q_n^{(0)} = M$, (4.47) yields

$$\Delta_n = \frac{1}{D}[M - (1-d)^3 Q_n^{(3)}] \tag{4.52}$$

$\{Q_n^{(3)}\}$ is periodic with period $2D$, thus we cannot say what will be the result of summing Δ_n over D clocks. We note, however that, since $\{Q_n^{(3)}\}$ are uniformly distributed on $[0, D-1]$, (4.52) implies that, for any integer K, the average value of Δ_n is

$$\frac{1}{K}\sum_{n=1}^{K} \Delta_n = \frac{M}{D} + O\left(\frac{1}{K}\right) \to \frac{M}{D}, \ K \gg 1 \tag{4.53}$$

Moreover, since $\{Q_n^{(3)}\}$ is periodic with period $2D$,

$$\sum_{n=1}^{2D} Q_{n-p}^{(3)} = \sum_{n=1}^{2D} Q_{n-q}^{(3)}, \ p, q = 0, 1, \ldots, 2D-1 \tag{4.54}$$

Thus, (4.52) yields

$$\sum_{n=1}^{2D} \Delta_n = 2M \tag{4.55}$$

It follows that the average value of Δ_n over one count pattern period $2D$, is exactly M/D as before.

Referring to (4.5), during each period of $2D$ reference clocks, the nth instantaneous error pulse $e_n(t)$ can be written, up to a constant multiplier, in the form

$$e_n(t) = u(t - nT_R) - u(t - nT_R - E_n)$$

$$E_n = \left(nN + \sum_{m=1}^{n} \Delta_m\right) T_0 - nT_R, \ n = 1, 2, \ldots, 2D \tag{4.56}$$

where E_n is the cumulative time error after n pulses, and $u(t)$ is the Heaviside function

$$u(t) = \begin{cases} 1, & t \geq 0 \\ 0, & t < 0 \end{cases} \tag{4.57}$$

After a full $2D$ period, we see from (4.56) that the cumulative time error E_{2D} vanishes if

$$\left(2DN + \sum_{m=1}^{2D} \Delta_m\right) T_0 - 2DT_R = 0 \tag{4.58}$$

Substituting (4.55) into (4.58) leads again to the division ratio (4.9)

$$(2DN + 2M)T_0 - 2DT_R = 0 \Rightarrow T_R = \left(N + \frac{M}{D}\right) T_0 \tag{4.59}$$

We note that E_n can have both positive or negative sign, and $|E_n|$ is the width of the periodic pulse $e_n(t)$ generated by the phase detector.

Under locking conditions, substituting T_R from (4.59) into (4.56) we may write

$$E_n = \left(\sum_{m=1}^{n} \Delta_m - n\frac{M}{D}\right) T_0, \quad n = 1, 2, \ldots, 2D \tag{4.60}$$

Substituting (4.52) into (4.60) and taking arbitrarily $Q_0^{(3)} = 0$, E_n turns out to be proportional to the second-order backward difference of $\{Q_n^{(3)}\}$, indeed

$$E_n = -\frac{T_o}{D} \sum_{m=1}^{n} (1-d)^3 Q_m^{(3)} = -\frac{T_o}{D}(1-d)^2 \sum_{m=1}^{n} (1-d)Q_m^{(3)}$$

$$= -\frac{T_o}{D}(1-d)^2 Q_n^{(3)} \tag{4.61}$$

Since usually $T_0 \ll T_R$, and since $|(1-d)^2 Q_n^{(3)}| < 2D$, then $|E_n| \ll T_R$. In this case, it has been noted in Perrott and Trott (2002) that it is reasonable to approximate, in the distribution sense (and normalizing dimensions)

$$e_n(t) \approx E_n \delta(t - nT_R) \tag{4.62}$$

and therefore the time error signal has the form

$$e(t) = \sum_{n=-\infty}^{\infty} e_n(t) \approx \sum_{n=-\infty}^{\infty} E_n \delta(t - nT_R) \tag{4.63}$$

To see that (4.62) holds, we compute the complex Fourier series coefficients c_k of $e(t)$ getting

$$c_k = \int_{0}^{(2DT_R)^-} \sum_{n=-\infty}^{\infty} [u(t - nT_R) - u(t - nT_R - E_n)]e^{-j\frac{\pi k}{DT_R}t} \, dt$$

$$= \sum_{n=0}^{2D-1} \int_{nT_R}^{nT_R+E_n} e^{-j\frac{\pi k}{DT_R}t} \, dt = \sum_{n=0}^{2D-1} \int_{nT_R}^{nT_R+E_n} e^{-j\frac{\pi k}{DT_R}t} \, dt$$

$$= \sum_{n=0}^{2D-1} \frac{DT_R}{-j\pi k} e^{-j\frac{\pi k}{D}n} \left(e^{-j\frac{\pi k}{D}\left(\frac{E_n}{T_R}\right)} - 1 \right) \approx \sum_{n=0}^{2D-1} E_n e^{-j\frac{\pi k}{D}n}$$

$$\approx \int_0^{(2DT_R)^-} \sum_{n=-\infty}^{\infty} E_n \delta(t - nT_R) e^{-j\frac{\pi k}{DT_R}t}\, dt, \quad \frac{|E_n|}{T_R} \ll 1 \tag{4.64}$$

where we used the approximation $e^{-x} \approx 1 - x, x \ll 1$.

Equation (4.64) shows that the Fourier series coefficients are approximately the same, using either (4.56) or (4.62) for $e_n(t)$.

For D large, and due to the complex relationships arising from the modulo-D addition, we may consider $\{Q_n^{(3)}, Q_m^{(3)}\}_{n \neq m}$ to be statistically independent random variables.

We show now that $\{Q_n^{(3)}\}$ are uniformly distributed on $[0, D-1]$. More generally, we show that if the input to a modulo-D adder is uniformly distributed on $[0, D-1]$, so is its output.

From (4.48)

$$Q_n^{(3)} = [Q_n^{(2)} + Q_{n-1}^{(3)}]_D$$

thus for $0 \leq k \leq D - 1$

$$P(Q_n^{(3)} = k) = P([Q_n^{(2)} + Q_{n-1}^{(3)}]_D = k)$$

$$= P(Q_n^{(2)} + Q_{n-1}^{(3)} = k) + P(Q_n^{(2)} + Q_{n-1}^{(3)} = D + k)$$

$$= \sum_{\alpha=0}^{k} P(Q_n^{(2)} = k - \alpha) P(Q_{n-1}^{(3)} = \alpha)$$

$$+ \sum_{\beta=0}^{D+k} P(Q_n^{(2)} = D + k - \beta) P(Q_{n-1}^{(3)} = \beta) \tag{4.65}$$

but, since $0 \leq Q_n^{(2)}, Q_{n-1}^{(3)} \leq D - 1$, then

$$P(Q_n^{(2)} = D + k - \beta) \neq 0 \Rightarrow k + 1 \leq \beta \leq D + k$$

$$P(Q_{n-1}^{(3)} = \beta) \neq 0 \Rightarrow \beta \leq D - 1 \tag{4.66}$$

From (4.42) it follows that $\{Q_n^{(2)}\}$ are uniformly distributed on $[0, D-1]$, then

$$P(Q_n^{(2)} = k) = \frac{1}{D}, \quad 0 \leq k \leq D - 1 \tag{4.67}$$

Substituting (4.66) and (4.67) into (4.65) we get

$$P(Q_n^{(3)} = k) = \frac{1}{D} \sum_{\alpha=0}^{k} P(Q_{n-1}^{(3)} = \alpha) + \frac{1}{D} \sum_{\beta=k+1}^{D-1} P(Q_{n-1}^{(3)} = \beta)$$

$$= \frac{1}{D} \sum_{\alpha=0}^{D-1} P(Q_{n-1}^{(3)} = \alpha) = \frac{1}{D} \tag{4.68}$$

Armed with (4.68) and denoting by $E[X]$ the expected value of X we may compute the mean and variance of $\{Q_n^{(3)}\}$

$$E[Q_n^{(3)}] = \frac{1}{D} \sum_{n=0}^{D-1} n = \frac{D-1}{2}$$

$$E[(Q_n^{(3)})^2] = \frac{1}{D} \sum_{n=0}^{D-1} n^2 = \frac{(D-1)(2D-1)}{6} \tag{4.69}$$

and

$$\sigma_Q^2 \equiv E[(Q_n^{(3)})^2] - E^2[Q_n^{(3)}]) = \frac{(D-1)(2D-1)}{6} - \frac{(D-1)^2}{4} = \frac{D^2-1}{12} \tag{4.70}$$

With c_k from (4.64) we compute the expected value of $|c_k|^2$

$$E[|c_k|^2] = E\left[\sum_{n=0}^{2D-1} E_n e^{-j\frac{\pi k}{D}n} \sum_{m=0}^{2D-1} E_m e^{j\frac{\pi k}{D}m} \right]$$

$$= \sum_{n=0}^{2D-1} \sum_{\substack{m=0 \\ |m-n|>2}}^{2D-1} E[E_n E_m] e^{-j\frac{\pi k}{D}(n-m)} + \sum_{n=0}^{2D-1} \sum_{p=-2}^{2} E[E_n E_{n+p}] e^{j\frac{\pi k}{D}p} \tag{4.71}$$

Now from (4.61)

$$E_n = -\frac{T_0}{D}(Q_n^{(3)} - 2Q_{n-1}^{(3)} + Q_{n-2}^{(3)}) \tag{4.72}$$

and since $\{Q_n^{(3)}, Q_m^{(3)}\}_{n \neq m}$ are statistically independent, it is immediate to verify that

$$E[E_n E_m] = 0, \ |n - m| > 2 \tag{4.73}$$

which reduces (4.71) to the form

$$E[|c_k|^2] = \sum_{n=0}^{2D-1} \sum_{p=-2}^{2} E[E_n E_{n+p}] e^{j\frac{\pi k}{D}p} \tag{4.74}$$

In view of (4.70) and (4.72), it is also straightforward to verify that, for large values of D

$$E[E_n E_{n+p}] = \begin{cases} 6\dfrac{T_0^2}{D^2}\sigma_Q^2 \approx \dfrac{T_0^2}{2}, & p = 0 \\[2mm] -4\dfrac{T_0^2}{D^2}\sigma_Q^2 \approx -\dfrac{T_0^2}{3}, & |p| = 1 \\[2mm] \dfrac{T_0^2}{D^2}\sigma_Q^2 \approx \dfrac{T_0^2}{12}, & |p| = 2 \\[2mm] & n = 0, 1, \ldots, D-1 \end{cases} \tag{4.75}$$

Using (4.75), Equation (4.74) becomes

$$E[|c_k|^2] = \frac{1}{3}DT_0^2\left[3 - 4\cos\left(\frac{\pi k}{D}\right) + \cos\left(\frac{2\pi k}{D}\right)\right], k = 0, 1, 2, \ldots \quad (4.76)$$

If we expand $\cos(\pi k/D)$, $\cos(2\pi k/D)$ in a Taylor series for small values of $2\pi k/D$, which is the relevant case if the error is filtered by the (slow) loop filter, the first non-canceling element in (4.76) is $DT_0^2(\pi k/D)^4/6$, thus

$$E[|c_k|^2] \approx \frac{1}{6}DT_0^2\left(\frac{\pi k}{D}\right)^4, \quad \frac{2\pi k}{D} \ll 1, \quad (4.77)$$

exhibiting a 40 dB/decade noise-shaping slope. Using the Fourier series for $e(t)$ we get

$$E[\|e(t)\|^2] = \left(\frac{1}{2DT_R}\right)^2 \sum_{k=-\infty}^{\infty} E[|c_k|^2] \approx \sum_{k \ll D/2\pi} \hat{f}_k \Delta f \quad (4.78)$$

We identify $\hat{f}_k$ in (4.78) as the discrete equivalent of the spectral representation of a random process, where, from (4.77)

$$\hat{f}_k = \frac{T_0^2}{12T_R}\left(\frac{\pi k}{D}\right)^4 \quad (4.79)$$

is integrated with the equivalent discrete integration step

$$\Delta f = \frac{1}{2DT_R} \quad (4.80)$$

We see that $\hat{f}_k$ has dimension [1/Hz], and is the approximate 'spectral density' of $e(t)$, which we will see shortly corresponds to the second-order derivative of a 'white noise' like signal.

If we use a first-order sigma-delta, then (4.61) yields

$$E_n = -(T_0/D)Q_n^{(1)}$$

Since $\{Q_n^{(1)}\}$ are uniformly distributed and independent, for large D

$$E[|c_k|^2] = \frac{T_0^2}{D^2}E\left[\sum_{n=0}^{2D-1} Q_n^{(1)}e^{-j\frac{\pi k}{D}n}\sum_{m=0}^{2D-1} Q_m^{(1)}e^{j\frac{\pi k}{D}m}\right]$$

$$= \frac{T_0^2}{D^2}\sum_{n=0}^{2D-1}\sum_{m\neq n}^{2D-1} E[Q_n^{(1)}Q_m^{(1)}]e^{-j\frac{\pi k}{D}(n-m)} + \frac{T_0^2}{D^2}2DE[(Q_n^{(1)})^2]$$

$$= \frac{T_0^2}{D^2}E^2[Q_n^{(1)}]\left\{\sum_{n=0}^{2D-1}\left(e^{-j\frac{\pi k}{D}n}\sum_{m=0}^{2D-1}e^{j\frac{\pi k}{D}m}\right) - 2D\right\} + \frac{2T_0^2}{D}E[(Q_n^{(1)})^2]$$

$$= \frac{2T_0^2}{D}\sigma_Q^2 + \frac{T_o^2}{D^2}\left(\frac{D-1}{2}\right)^2 (4D^2 - 2D)\delta_{k,0}$$

$$\approx \frac{1}{6}DT_0^2 + D^2 T_0^2 \delta_{k,0} \qquad (4.81)$$

and therefore, for $k \le K < \infty$ we get a 'white-noise' error

$$E[\|e(t)\|^2] = \left(\frac{1}{2DT_R}\right)^2 \sum_{k=-K}^{K} E[|c_k|^2]$$

$$\approx \frac{T_0^2}{4T_R^2} + \sum_{\substack{k=-K \\ k \ne 0}}^{K} \hat{f}_{white}\Delta f, \ \hat{f}_{white} = \frac{T_0^2}{12T_R}, \Delta f = \frac{1}{2DT_R} \quad (4.82)$$

We see that $\hat{f}_k$ in (4.79) is the 'spectral density' $\hat{f}_{white}$ of an uniformly distributed and statistically independent sequence such as $\{Q_n^{(3)}\}$, similar in characteristics to (4.82), but multiplied by $|\pi k/D|^4$ in (4.79). Thus, taking the second-order difference of $\{Q_n^{(3)}\}$ is the discrete equivalent of differentiating twice a 'bandlimited-white' random noise signal, thus multiplying its spectral density by $|\omega|^4$.

If $k \ll D/2\pi$ then $(\pi k/D)^4 \ll 1$, thus, comparing (4.78) and (4.80) to (4.82), we see that the total low-frequency steering-line noise of the third-order MASH synthesizer is substantially reduced as compared to the first-order case (neglecting DC) if

$$\left(\frac{\pi}{D}\right)^4 \sum_{k=1}^{K} k^4 \approx \left(\frac{\pi}{D}\right)^4 \frac{K^5}{5} \ll K \Rightarrow K \ll \frac{D}{2} \qquad (4.83)$$

which, in other words, means that if we denote by f_{LF} the corner frequency of the PLL loop filter, (4.83) requires

$$f_{LF} = K\Delta f = \frac{K}{2D} f_R \ll \frac{1}{4} f_R \qquad (4.84)$$

and then from (4.83) the noise improvement NI is

$$NI \approx 40 \log_{10}(f_R/4f_{LF}) = 40 \log_{10}(D/2K) \ [dB], \ f_R/4f_{LF} \gg 1 \quad (4.85)$$

Figure 4.4(a) and (b) show the simulation of the phase detector noise density for first-order sigma-delta, and the third-order MASH implementation with

$$D = 2048, \ M = 1 \text{ and } Q_0^{(1)} = Q_0^{(2)} = Q_0^{(3)} = 0$$

Figure 4.5 shows a log-log plot of Figure 4.4(b). Each plot shows the noise power as a function of frequency, normalized to the total noise power. It should

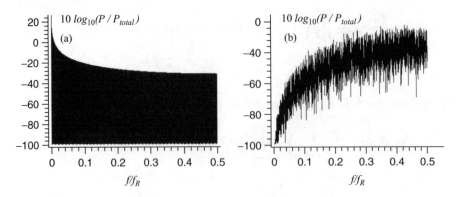

Figure 4.4 (a) Spectral density for the first-order sigma-delta with $M = 1$. (b) Spectral density for the third-order MASH with $M = 1$

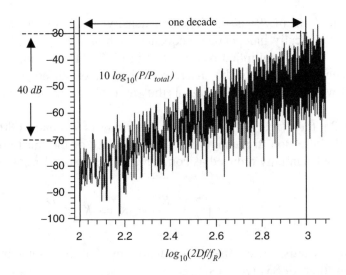

Figure 4.5 Log-log plot of Figure 4.4(b) showing 40 dB/decade noise shaping

be noted that the first-order sigma-delta has non-zero 'DC' and 'white' spectral shape. In contrast, the third-order MASH has zero mean (no DC, which simplifies the phase detector implementation), and its noise is shaped with a 40 dB/decade slope.

4.5.1 Pseudorandom Sequence Bounds

There is one additional important property of the MASH pseudorandom counting sequence $\{\Delta_n\}$ that is worth attention. Denote by Δ_n^N the nth incremental counting value in a MASH architecture of order N, then

$$-(2^{N-1} - 1) \leq \Delta_n^N \leq 2^{N-1} \tag{4.86}$$

We prove (4.86) by induction on N. First, let us rewrite (4.47) for the general case

$$\Delta_n^N = \sum_{k=1}^{N} (1-d)^{k-1} c_n^k, \quad c_n^k \in \{0, 1\} \tag{4.87}$$

For $N = 1$, (4.86) is satisfied, since (4.87) implies that $\Delta_n^1 = c_n^1 \in \{0, 1\}$. We now take (4.86) to be the induction assumption of order N, then

$$\Delta_n^{N+1} = \sum_{k=1}^{N+1} (1-d)^{k-1} c_n^k = \sum_{k=1}^{N} (1-d)^{k-1} c_n^k + (1-d)^N c_n^{N+1} \tag{4.88}$$

Substituting (4.87) into (4.88) and using the induction assumption (4.86)

$$-(2^{N-1} - 1) + (1-d)^N c_n^{N+1} \le \Delta_n^{N+1} \le 2^{N-1} + (1-d)^N c_n^{N+1} \tag{4.89}$$

We prove the right-hand side of (4.89). The proof for the left side is identical. Let us expand the delay operator in binomial form

$$(1-d)^N c_n^{N+1} = \left[\sum_{m=0}^{N} \binom{N}{m} (-1)^m d^m \right] c_n^{N+1} \tag{4.90}$$

The value of (4.90) will be maximal when all the carry bits that experienced an even number of delays (and therefore have positive coefficients) are '1', and all the carry bits that experienced an odd number of delays (and therefore have negative coefficients) are '0'. Therefore, taking only the even elements in the sum, (4.90) yields

$$(1-d)^N c_n^{N+1} \le \sum_{m=0}^{[N/2]} \binom{N}{2m} \tag{4.91}$$

where $[N/2]$ denotes the integer value of $N/2$. Now, we recall the property of the binomial coefficients

$$\binom{N}{0} + \binom{N}{2} + \binom{N}{4} + \cdots = \binom{N}{1} + \binom{N}{3} + \binom{N}{5} + \cdots = 2^{N-1} \tag{4.92}$$

which by (4.91) implies

$$(1-d)^N c_n^{N+1} \le 2^{N-1} \tag{4.93}$$

We complete the proof by substituting (4.93) into (4.89) getting

$$\Delta_n^{N+1} \le 2^{N-1} + 2^{N-1} = 2^N \tag{4.94}$$

Figure 4.6(a) shows a plot of the sequence $\{\Delta_n\}$ for the third-order MASH architecture of Figure 4.3 with $D = 256$, $M = 151$ and all the adders cleared at counting start. Figure 4.6(b) shows a close up of Figure 4.6(a).

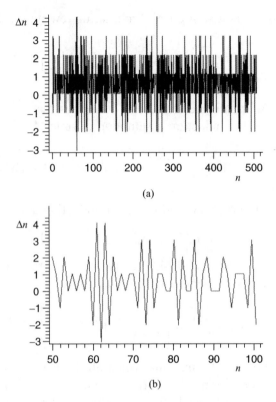

(a)

(b)

Figure 4.6 Plots of the incremental pseudorandom sequence $\{\Delta_n\}$

4.6 Analog Sigma-Delta A/D Converter

This is an appropriate point to analyze the ancestor of the digital sigma-delta, namely the analog sigma-delta modulator, widely used as a fast high-resolution A/D converter for sampling the I and Q baseband channels.

Referring to Figure 4.7, let $f(t)$ be an analog signal of bandwidth much smaller than the acquisition frequency $1/T$, and with $|f(t)| < K$.

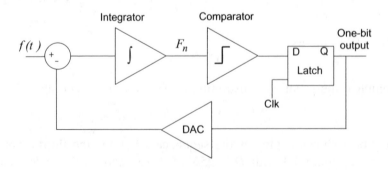

Figure 4.7 Analog sigma-delta modulator

Denoting the clock period by $\tau \ll T$, let F_n be the value of the output of the integrator at the instant $t_0 + n\tau$ at which the nth clock occurs. Assuming that the comparator outputs a logic '1' if $F_n \geq 0$ or a logic '0' if $F_n < 0$, the DAC outputs $+K$ or $-K$ in correspondence to $Q = 1$ or $Q = 0$ respectively, and with $|F_1| < \infty$, we may write

$$F_{n+1} = F_n + \int_{n\tau}^{(n+1)\tau} [f(t) - K\,sign(F_n)]\,dt, \quad n = 1, 2, 3\ldots, |F_1| < \infty \quad (4.95)$$

Denoting by $[t_0, t_0 + T]$ the acquisition interval, we may expand $f(t)$ in a first-order Taylor series

$$f(t) = f(t_n) + (t - t_n)f'(\xi(t)), \xi(t) \in [t_n, t_n + \tau], t_n = n\tau \quad (4.96)$$

By the mean-value theorem for integrals, we may write

$$\left| \int_{n\tau}^{(n+1)\tau} (t - t_n)f'(\xi(t))\,dt \right| \leq \tau^2|f'(\xi_n)| \equiv e_n, \xi_n \in [t_n, t_n + \tau] \quad (4.97)$$

from which (4.95) yields

$$F_{n+1} = F_n + \tau[f(t_n) - K\,sign(F_n)] + O(e_n) \quad (4.98)$$

Now note that since $|f(t)| < K$, then the sequence $\{F_n\}$ is bounded, for if $F_n \geq 0$ then the input to the integrator will be negative, thus $F_{n+1} < F_n$, and similarly, if $F_n < 0$ the input to the integrator will be positive, thus $F_{n+1} > F_n$. Therefore, no matter where it started, $\{F_n\}$ will end up oscillating near zero with alternating sign.

It follows from (4.95) that, since the sign of the integrand is opposite to the sign of F_n, then, after the first zero-crossing, $|F_{n+1}|$ will reach its maximal value if $F_n \in \{0^+, 0^-\}$, and

$$|F_{n+1}| < 2K\tau \quad (4.99)$$

Since $\{F_n\}$ is bounded, then

$$|F_j| < \infty \Rightarrow |F_n - F_m| < \infty, n, m, j \in \{1, 2, 3\ldots\} \quad (4.100)$$

and in particular $|F_{n+1} - F_1| < \infty$, thus

$$\lim_{n\to\infty} \frac{1}{n}\left|F_{n+1} - F_1\right| = \lim_{n\to\infty} \frac{1}{n}\left|\sum_{k=1}^{n}(F_{k+1} - F_k)\right| = 0 \quad (4.101)$$

and using (4.98) we may write

$$0 = \lim_{n\to\infty} \frac{1}{n}|(F_{n+1} - F_n) + (F_n - F_{n-1}) + \cdots + (F_2 - F_1)|$$

$$= \lim_{n \to \infty} \frac{1}{n} \left| \tau \sum_{q=1}^{n} [f(t_q) - K \, sign(F_q)] + \sum_{q=1}^{n} O(e_q) \right|$$

$$= \lim_{n \to \infty} \left| \tau \left[f_{ave} - K \frac{1}{n} \sum_{q=1}^{n} sign(F_q) \right] + O \left(\tau^2 \frac{1}{n} \sum_{q=1}^{n} |f'(\xi_q)| \right) \right| \quad (4.102)$$

where f_{ave} is the average value over the acquisition time. Therefore, if f is nearly constant during sampling, (4.102) implies

$$f(t_0) \approx f_{ave} = \lim_{n \to \infty} \left[K \frac{1}{n} \sum_{q=1}^{n} sign(F_q) - O \left(\frac{1}{n} \sum_{q=1}^{n} \tau |f'(\xi_q)| \right) \right] \quad (4.103)$$

For n finite, and denoting by T the A/D acquisition time, namely

$$1 \le n \le N, T = N\tau \quad (4.104)$$

Equation (4.102) yields a modification of (4.103) in the form

$$f(t_0) = K \frac{1}{N} \sum_{q=1}^{N} sign(F_q) - O \left(\frac{1}{N} \sum_{q=1}^{N} \tau |f'(\xi_q)| \right) \quad (4.105)$$

where the order of magnitude of the error satisfies

$$O \left(\frac{1}{N} \sum_{q=1}^{N} \tau |f'(\xi_q)| \right) \le O \left(\frac{1}{N} T \max_n |f'(\xi_q)| \right) \quad (4.106)$$

If $f(t)$ is bandlimited with bandwidth $B \ll 1/T$, namely, $f(t)$ varies only little during the acquisition interval, then

$$T \max_n |f'(\xi_n)| \ll 2K \quad (4.107)$$

and therefore, for N large, we may approximate (4.105) to the form

$$f(t_0) \approx K \left[\frac{1}{N} \sum_{q=1}^{N} sign(F_q) + O(2/N) \right] \quad (4.108)$$

In other words, for any given acquisition interval $[t_0, t_0 + T]$, the approximation value for $f(t_0)$ is K times the average of the comparator signs, with a relative error of the order of magnitude of $2/N$.

If N is taken to be a power of two, say $N = 2^m$, all we need to do in order to obtain the A/D conversion value, is to attach an up/down counter to the clock,

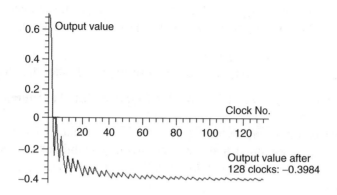

Figure 4.8   Convergence of the sigma-delta A/D converter

with the up/down control connected to the Q bit of the latch of Figure 4.7. Then denoting by $\{Q_0, Q_1, \ldots, Q_{D-1}\}$, the output word of the counter, the output word of the D/A is taken to be its mth left-shift $\{Q_{m+1}, Q_{m+2}, \ldots, Q_{D-1}0, \ldots, 0\}$, which is the same as the value obtained by dividing the result by N.

Figure 4.8 shows an exemplary simulation of the convergence of the sigma-delta A/D converter as described, with $K = 1$, $N = 128$, $F_1 = 0.67$, and $f(t_0) = -0.41$. The approximation error is 2.8% which is in good agreement with the order of magnitude $2/128 = 1.6\%$.

If we increase N about 10 times, namely $N = 1024$, then the output value becomes -0.40918, and the approximation error reduces to 0.2% in close agreement with $2/1024 = 0.195\%$.

4.7 Review of PLL Fundamentals

Among the many types of synthesizer architectures, we accurately review here the one which constitutes the most practical and commonly employed configuration in digital communication equipment, namely, the 'second-order, type-two phase locked loop' driven by a 'charge-pump' phase detector. The reasons of its popularity are simple construction and straightforward mathematical treatment, along with good performance and stability characteristics.

It should be pointed out, however, that, in general, a 'pure' second-order type-two PLL cannot deliver the required reference spur characteristics, and some modifications, in the form of extra attenuation, must be added in the loop filter.

Usually, the extra attenuation can be neglected in operational computations, but it results in the introduction of parasitic poles, generating in fact a potentially unstable third-order transfer function that may lead to PLL oscillations.

In the analysis to follow we neglect the parasitic poles, and we defer their treatment to Chapter 6, where we show how to simply and accurately evaluate their effect by means of 'perturbation methods'.

For the sake of easy understanding, we first analyze the integer-N case, and then we explain how to apply the results to the fractional-N context.

4.7.1 Basic Integer-N Configuration

The basic integer-N PLL configuration is described in Figure 4.9, where

- ω_{ref} [rad/sec] denotes the fixed angular frequency of the reference oscillator signal $\omega_{ref} = 2\pi f_{ref}$.
- ω_0 [rad/sec] is a constant denoting the angular frequency of the VCO at time $t = 0$, where it is assumed that the synthesizer is stabilized and locked at the fixed frequency ω_0 under the condition $\Delta N = 0$. Of course, ω_0 depends (among others) on N.
- ΔN is an integer denoting a change of the programmable divider value from N to $N + \Delta N$ occurring at the time $t = 0$, and causing the VCO to leave the present steady-state condition and 'jump' to a new frequency.
- $h(t)$ [volt/amp] denotes the (time) impulse response of the loop filter, which in general is a low-pass filter of bandwidth B [Hz]. All the low-frequency signals in the loop are sampled and ultimately filtered by $h(t)$. The loop filter accepts a current at its input and delivers a voltage at its output.
- $\Delta\omega(t)$ [rad/sec] denotes the instantaneous frequency offset of the VCO output signal $f_o(t)$ with respect to the previous steady-state value ω_0. Clearly, $\Delta\omega(t) = 0$ for $t \leq 0$. Thus, $\Delta\omega(t)$ describes the behavior of $f_o(t)$ for $t > 0$, during the transition from the steady-state condition for $t = 0$ and $\Delta N = 0$ to the new steady-state condition for $\Delta N \neq 0$ and $t \to \infty$

$$f_o(t) = cos\left[\omega_0 t + \int_0^t \Delta\omega(t)\,dt\right]$$

- $\theta_{out}(t)$ [rad] denotes the instantaneous phase of the output signal after being divided by the programmable divider.

$$\theta_{out}(t) = \frac{1}{N + \Delta N}\left(\omega_0 t + \int_0^t \Delta\omega(t)\,dt\right)$$

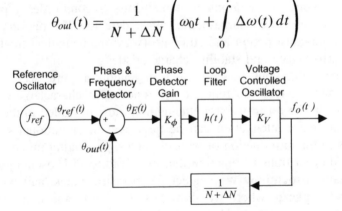

Figure 4.9   Basic integer-N configuration

- $\theta_{ref}(t)$ [rad] denotes the instantaneous phase of the reference oscillator signal

$$\theta_{ref}(t) = \omega_{ref}t$$

- $\theta_E(t)$ [rad] denotes the instantaneous phase error of the divided output signal with respect to the phase of the reference signal

$$\theta_E(t) = \theta_{ref}(t) - \theta_{out}(t)$$

- $K\phi$ [amp/rad] is a constant denoting the gain of the 'charge-pump' type phase detector. This detector type accepts two square waves at its input, and outputs a series of current pulses, at reference rate, whose average value is directly proportional to their phase difference modulo 2π. The phase detector begins to work when the cumulative phase error between divided VCO and reference signals is within $\pm 2\pi$. Under this condition the two signals are considered at the same frequency. If the cumulative phase difference exceeds $\pm 2\pi$, then the two signals are interpreted to be at different frequencies, and the detector automatically switches to a mode called 'frequency mode', outputting a current with polarity that tends to correct the frequency difference. In a properly designed system, the 'frequency mode' is active only during start up, and does not come to activation during subsequent frequency jumps, thus, for computation purposes, the detector is always in 'phase mode'. For the sake of completeness, we include a brief description of the phase-frequency detector operation at the end of this review.
- K_V [(rad/sec)/volt] is a constant denoting the 'gain' of the VCO. The VCO accepts the loop filter voltage into the steering line (Chapter 5), and outputs an angular frequency change directly proportional to that voltage.

NOTE: It should be noted that the output of the phase detector is sampled every reference pulse, namely at frequency f_{ref}, thus, the sampling interval for the phase error $\theta_E(t)$ is $T_s = 1/f_{ref}$. It follows that the bandwidth of the loop filter should be limited to $B < f_{ref}/2$, else aliasing will occur. In practice, in order to obtain reasonable reference spurs, we usually take $B \ll f_{ref}$. Typically $B < f_{ref}/10$. With the above approximation in place, we may disregard the fact that the phase detector works in sampling mode, and we may carry on the computations as if we were treating analog signals.

In analyzing synthesizer characteristics when performing a 'frequency jump', it is customary to compute the PLL behavior following a step change in the reference frequency. This nearly equivalent representation somewhat simplifies the analysis, but does not reflect the actual process. In reality, the reference frequency is fixed, and the frequency jump is obtained by reprogramming the programmable counter at the instant $t = 0$ from some present value N to a new value $N + \Delta N$. For the sake of clarity and understanding, this last scenario is the one we will use in our analysis.

4.7.2 Integer-N Transient Analysis

For $t < 0$, we assume that the PLL is locked under steady-state conditions, namely the divided VCO phase is identical to the reference phase, and thus

$$\Delta\omega(t) = 0, \quad \theta_E(t) = 0, \quad \Delta N = 0$$

$$f_o(t) = \cos(\omega_0 t), \quad \omega_0/N = \omega_{ref}, \quad t < 0 \tag{4.109}$$

At the instant $t = 0$ we reprogram the counter so that $N \to N + \Delta N$, namely

$$\Delta N \neq 0, t \geq 0 \tag{4.110}$$

and we look to see how the above change affects $\Delta\omega(t)$ for $0 \leq t \leq \infty$.

Since the instantaneous phase is the integral of the instantaneous frequency, we may write

$$\theta_{ref}(t) - \theta_{ref}(0) = \int_0^t \omega_{ref}\, d\xi = \omega_{ref}\, t$$

$$\theta_{out}(t) - \theta_{out}(0) = \int_0^t \frac{\omega_0 + \Delta\omega(\xi)}{N + \Delta N}\, d\xi \tag{4.111}$$

Assuming that $\Delta N \ll N$, which is always the case, we may approximate

$$\frac{1}{N + \Delta N} \approx \frac{1}{N}\left(1 - \frac{\Delta N}{N}\right) \tag{4.112}$$

Assuming initial phase coherence at transition start, we set

$$\theta_{ref}(0) = \theta_{out}(0) = 0 \tag{4.113}$$

and with $\theta_E(t) = \theta_{ref}(t) - \theta_{out}(t)$, (4.111) yields

$$\theta_E(t) \approx \omega_{ref}\, t \cdot u(t) - \frac{\omega_0 t}{N} u(t) + \frac{\Delta N}{N}\frac{\omega_0 t}{N} u(t) - \int_0^t \frac{\Delta\omega(\xi)}{N}\left(1 - \frac{\Delta N}{N}\right) d\xi \tag{4.114}$$

where $u(t)$ is the Heaviside function. Recalling that

$$\frac{\omega_0}{N} = \omega_{ref}, \quad \frac{\Delta N}{N} \ll 1 \tag{4.115}$$

then (4.114) may be further approximated to the form

$$\theta_E(t) = \frac{\Delta N}{N}\omega_{ref}\, t \cdot u(t) - \int_0^t \frac{\Delta\omega(\xi)}{N}\, d\xi \tag{4.116}$$

Looking at the diagram of Figure 4.9 we may derive the equation

$$K_\phi K_V [\theta_E * h](t) = \Delta\omega(t) \tag{4.117}$$

where $[\theta_E * h](t)$ denotes time convolution

$$[\theta_E * h](t) = \int_{-\infty}^{\infty} \theta_E(\tau) h(t - \tau)\, d\tau \tag{4.118}$$

Equations (4.116) and (4.117) form a set of integral equations whose solution is easily found in the Laplace transform domain. Denoting the Laplace transform of $f(t)$ by $F(s)$

$$F(s) = L[f(t)] = \int_0^\infty f(t) e^{-st}\, dt, \quad s = \sigma + j\omega, \sigma, \omega \in \mathbb{R} \tag{4.119}$$

where s is the complex-valued variable in the transform domain, we may rewrite the set (4.116) and (4.117) in the form

$$\Theta_E(s) = \frac{\Delta N}{Ns^2}\omega_{ref} - \frac{\Delta\Omega(s)}{Ns}$$

$$K_\phi K_V \Theta_E(s) H(s) = \Delta\Omega(s) \tag{4.120}$$

where we make use of the Laplace pairs

$$L\left[\int_0^t f(\xi)\, d\xi\right] = \frac{1}{s} F(s), \quad L[t^n \cdot u(t)] = \frac{n!}{s^{n+1}}$$

$$L[\theta_E * h](t) = \Theta_E(s) H(s) \tag{4.121}$$

Assuming that $\theta_E(t)$ in (4.116) is bounded when $t \to \infty$ (we verify later the validity of that), and eliminating $\Theta_E(s)$ from the set (4.120) we obtain

$$\Delta\Omega(s) = \frac{1}{s} \frac{K_\phi K_V H(s)}{Ns + K_\phi K_V H(s)} \Delta N\omega_{ref} \tag{4.122}$$

Applying to (4.122) the final-value theorem of Laplace transform, namely

$$\lim_{t\to\infty} f(t) = \lim_{s\to0} s F(s) \tag{4.123}$$

we obtain the steady-state output frequency change due to the change ΔN

$$\lim_{t\to\infty} \Delta\omega(t) = \lim_{s\to0} s\Delta\Omega(s)$$

$$= \lim_{s\to0} \frac{K_\phi K_V H(s)}{Ns + K_\phi K_V H(s)} \Delta N\omega_{ref} = \Delta N\omega_{ref} \tag{4.124}$$

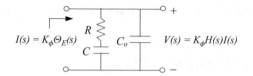

Figure 4.10   Loop filter topology

We see that the loop stabilizes at the new frequency value $\omega_0 + \Delta N \omega_{ref}$ independently from the loop characteristics.

The issue of interest, however, is how fast and how stably $\Delta \omega(t)$ approaches $\Delta N \omega_{ref}$. The answer to this question depends on $h(t)$, N, K_ϕ and K_V. Since K_ϕ and K_V are constants and may be incorporated into $h(t)$, and N is dictated by the required steady-state frequency, then $h(t)$ is the key element in the design.

Here we consider only loop filters whose transfer function $H(s)$ can be approximated to the form

$$H(s) = K_L \frac{s + \omega_L}{s} \tag{4.125}$$

In particular such a common filter has the topology shown in Figure 4.10. If we assume that $C_0 \ll C$, we may set

$$C_0 = \varepsilon C, \quad \varepsilon \to 0 \tag{4.126}$$

so that the transfer function of the loop filter of Figure 4.10 has the form

$$
\begin{aligned}
H(s) &= K_L \frac{s + \omega_L}{s} \cdot \frac{1}{\dfrac{\varepsilon}{\omega_L} s + 1 + \varepsilon} \\
&= K_L \frac{s + \omega_L}{s} \cdot \frac{\omega_p}{s + \omega_p(1 + \varepsilon)} \approx K_L \frac{s + \omega_L}{s}, \quad \varepsilon \to 0 \tag{4.127}
\end{aligned}
$$

where

$$K_L = R, \quad \omega_L = 1/RC, \quad \omega_p = \omega_L/\varepsilon \to \infty, \quad \varepsilon \to 0 \tag{4.128}$$

The purpose of the capacitor C_0, often referred to as the 'pre-integration capacitor', is to create an additional loop filter pole at about ω_p such that

$$\omega_L \ll \omega_p \ll 2\pi f_{ref} \tag{4.129}$$

The additional pole makes the loop filter act as a low-pass filter for the phase error signal which is sampled at rate f_{ref}, thus both preventing aliasing and reducing the reference spurs due to the pulses out of the phase detector. At the same time this additional 'parasitic' pole is high enough in frequency that it can be neglected for the purpose of lock-up time computations.

However, as mentioned before, the extra parasitic pole may jeopardize stability and cause PLL oscillations. For the moment we disregard this problem, and we defer its in-depth discussion to Chapter 6 which deals with parasitic effects.

Substituting (4.125) into (4.122) yields the explicit expression for the output frequency offset from ω_0 produced by a change ΔN, at the instant $t = 0$, from the steady state counting value N

$$\Delta\Omega(s) = \frac{\alpha s + \alpha\omega_L}{s(s^2 + \alpha s + \alpha\omega_L)}\Delta N\omega_{ref}$$

$$\alpha = \frac{K_\phi K_V K_L}{N} \in \mathbb{R} \tag{4.130}$$

It is customary to put the quadratic expression of (4.130) in the form

$$s^2 + \alpha s + \alpha\omega_L \equiv s^2 + 2\xi\omega_n s + \omega_n^2, \ \xi, \omega_n > 0 \tag{4.131}$$

where, for reasons to become clear shortly, ω_n is referred to as 'the natural frequency', and ξ 'the damping factor'. Equation (4.131) implies

$$\xi = \frac{1}{2}\sqrt{\frac{\alpha}{\omega_L}}, \ \omega_n = \sqrt{\alpha\omega_L} \tag{4.132}$$

Setting

$$s^2 + 2\xi\omega_n s + \omega_n^2 = (s - \gamma_1)(s - \gamma_2) \tag{4.133}$$

we may rewrite (4.130) in the form

$$\Delta\Omega(s) = \frac{-(\gamma_1 + \gamma_2)s + \gamma_1\gamma_2}{s(s - \gamma_1)(s - \gamma_2)}\Delta N\omega_{ref}$$

$$\gamma_1, \gamma_2 = \omega_n(-\xi \pm \sqrt{\xi^2 - 1}) \tag{4.134}$$

If $\gamma_1 \neq \gamma_2$, we may expand (4.134) in a partial fraction expansion of the form

$$\Delta\Omega(s) = \left(\frac{A}{s} + \frac{B}{s - \gamma_1} + \frac{C}{s - \gamma_2}\right)\Delta N\omega_{ref}, \ \gamma_1 \neq \gamma_2 \tag{4.135}$$

By recombining (4.135) and comparing the result with (4.134), we identify

$$A = 1, \ B = \frac{\gamma_1}{\gamma_2 - \gamma_1}, \ C = -\frac{\gamma_2}{\gamma_2 - \gamma_1}, \ \gamma_1 \neq \gamma_2 \tag{4.136}$$

If $\gamma_1 = \gamma_2 = \gamma$, then (4.134) becomes

$$\Delta\Omega(s) = \frac{-2\gamma s + \gamma^2}{s(s - \gamma)^2}\Delta N\omega_{ref}, \ \gamma = -\omega_n \tag{4.137}$$

and has the partial fraction expansion

$$\Delta\Omega(s) = \left(\frac{D}{s} + \frac{E}{s - \gamma} + \frac{F}{(s - \gamma)^2}\right)\Delta N\omega_{ref}, \ \gamma_1 = \gamma_2 = \gamma \tag{4.138}$$

yielding

$$D = 1, \quad E = -1, \quad F = -\gamma, \quad \gamma_1 = \gamma_2 = \gamma \tag{4.139}$$

Denoting the inverse Laplace transform of $F(s)$ by

$$f(t) = L^{-1}[F(s)] \tag{4.140}$$

we recall that

$$L^{-1}\left[\frac{1}{s+\lambda}\right] = e^{-\lambda t}u(t), \quad L^{-1}\left[\frac{1}{(s+\lambda)^2}\right] = te^{-\lambda t}u(t), \quad \lambda \in \mathbb{C} \tag{4.141}$$

Applying (4.141) to (4.135), (4.136) and (4.138), (4.139) yields

$$\Delta\omega(t) = \left(1 + \frac{\gamma_1}{\gamma_2 - \gamma_1}e^{\gamma_1 t} - \frac{\gamma_2}{\gamma_2 - \gamma_1}e^{\gamma_2 t}\right)u(t)\Delta N\omega_{ref}, \quad \gamma_1 \neq \gamma_2 \tag{4.142}$$

$$\Delta\omega(t) = (1 - e^{\gamma t} - \gamma te^{\gamma t})u(t)\Delta N\omega_{ref}, \quad \gamma_1 = \gamma_2 = \gamma \tag{4.143}$$

Equation (4.142) has two fundamentally different solutions depending whether $\xi < 1$ or $\xi \geq 1$. Indeed looking at (4.134), we can see that

- If $\xi < 1$ then the roots are complex conjugate and we denote them by $\gamma_1 = \gamma$ and $\gamma_2 = \overline{\gamma}$, where the upper bar notation indicates complex conjugate. This case is referred to as the 'under-damped case', and we show that it leads to an oscillatory (ringing) lock-up process of the PLL. If $\xi \to 0$, namely, in the extremely under-damped case, then $\gamma \to j\omega_n$ which, as we see shortly, approaches sinusoidal oscillations at frequency ω_n. This is why ω_n is referred to as the 'natural frequency'. The under-damped case is the one of most interest, as it is the one of most practical use.
- If $\xi = 1$ we have two identical roots (a root of multiplicity two). This case, which is a limit case of the under-damped condition, is referred to as the 'critically damped case'. Here the lock-up process is smooth, without ringing.
- If $\xi > 1$, we still have two distinct roots and no ringing, but this case, referred to as the 'over-damped case', is not of very practical use, and we will not investigate it. The interested reader may derive its behavior using the very same procedure as for the under-damped case.

Setting, for the under-damped case

$$\gamma_1 = \gamma, \gamma_2 = \overline{\gamma}, \gamma = a + jb = \omega_n(-\xi + j\sqrt{1 - \xi^2}) \tag{4.144}$$

since

$$\phi = tan^{-1}\frac{sin\,\phi}{cos\,\phi} = sin^{-1}(sin\,\phi) \tag{4.145}$$

and noting that

$$|\gamma|^2 = a^2 + b^2 = \omega_n^2 \xi^2 + \omega_n^2(1 - \xi^2) = \omega_n^2 \qquad (4.146)$$

we conclude that there exists an angle ϕ such that

$$\phi = tan^{-1}\frac{b}{a} = -tan^{-1}\frac{\sqrt{1-\xi^2}}{\xi} = -sin^{-1}(\sqrt{1-\xi^2}) \qquad (4.147)$$

With the help of (4.144) and (4.147), Equation (4.142) is rearranged in the form

$$
\begin{aligned}
\Delta\omega(t) &= \left(1 + \frac{a+jb}{j2b}e^{(a+jb)t} + \frac{a-jb}{j2b}e^{(a-jb)t}\right)u(t)\Delta N\omega_{ref} \\
&= \left(1 + \sqrt{a^2+b^2}e^{-at}\frac{e^{j[bt+tg^{-1}(b/a)]} - e^{-j[bt+tg^{-1}(b/a)]}}{j2b}\right)u(t)\Delta N\omega_{ref} \\
&= \left[1 + \frac{e^{-\xi\omega_n t}}{\sqrt{1-\xi^2}}\sin(\omega_n\sqrt{1-\xi^2}t - \sin^{-1}\sqrt{1-\xi^2})\right]u(t)\Delta N\omega_{ref}
\end{aligned}
$$

$$(4.148)$$

In the critically damped case we get from (4.134)

$$\gamma_1 = \gamma_2 = \gamma = -\omega_n \qquad (4.149)$$

and substituting (4.149) into (4.143) or taking the limit of (4.148) when $\xi \to 1$ yields

$$\Delta\omega(t) = [1 + e^{-\omega_n t}(\omega_n t - 1)]u(t)\Delta N\omega_{ref} \qquad (4.150)$$

It is customary to represent (4.148) and (4.150) in normalized form by plotting $\Delta\omega/\Delta N\omega_{ref}$ vs. $\omega_n t$ as in Figure 4.11.

Now, It is easy to see from (4.120) and (4.125) that when $t \to \infty$, the phase error at the phase detector vanishes (from which the name 'phase locked' loop arises). Indeed, isolating $\Theta_E(s)$ in (4.120) yields

$$\Theta_E(s) = \frac{1}{s}\frac{\Delta N\omega_{ref}}{Ns + K_\phi K_V H(s)} \qquad (4.151)$$

Substituting (4.125) into (4.151) and using the final value theorem (4.123), we get

$$\lim_{t\to\infty}\theta_E(t) = \lim_{s\to 0} s\Theta_E(s) = \lim_{s\to 0}\frac{\Delta N\omega_{ref}s}{Ns^2 + K_\phi K_V K_L(s + \omega_L)} = 0 \qquad (4.152)$$

Since most of the available digital phase detectors work in modulo 2π at most, it is important to verify that during the worst-case frequency jump, the instantaneous phase error satisfies $|\theta_E(t)| < 2\pi$, or the phase detector will temporarily enter

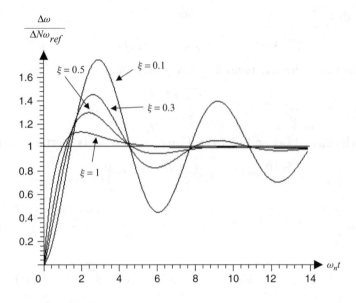

Figure 4.11   Normalized frequency transient

the 'frequency mode'. Although the detector will eventually return to the 'phase mode', such an event will jeopardize the PLL lock time.

Let us analyze the worst-case phase error during a frequency jump. The extreme points of the phase error are found by equating to zero the derivative of $\theta_E(t)$ in (4.116) for $t > 0$, namely after a jump start

$$\theta'_E(t) = \frac{\Delta N}{N}\omega_{ref} - \frac{\Delta\omega(t)}{N} = 0 \Rightarrow \Delta\omega(t) = \Delta N\omega_{ref} \qquad (4.153)$$

With reference to Figure 4.11, we can see that (4.153) is satisfied each time a plot crosses the horizontal line that intersects the value '1' on the vertical axis. In the under-damped case, substituting (4.153) in (4.148) the extreme points $\{t_k\}$ are for

$$\omega_n\sqrt{1 - \xi^2}t_k - sin^{-1}\sqrt{1 - \xi^2} = k\pi, \ k \in \mathbb{N} \qquad (4.154)$$

namely

$$t_k = \frac{k\pi + sin^{-1}\sqrt{1 - \xi^2}}{\omega_n\sqrt{1 - \xi^2}}, \ k \in \mathbb{N} \qquad (4.155)$$

Since, as shown in (4.152), the phase error tends to zero, we intuitively feel that the maximal instantaneous phase error during the lock-up process must occur at the first extreme point encountered. The mathematical formulation of this fact is obtained as follows: we substitute (4.148) in (4.116) getting

$$\theta_E(t) = \frac{\Delta N}{N}\omega_{ref}t \cdot u(t) - \int_0^t \frac{\Delta\omega(u)}{N} du = A\int_0^t f(u)\,du \qquad (4.156)$$

where

$$f(u) = e^{-\xi \omega_n u} \sin\left(\omega_n \sqrt{1-\xi^2}\, u - \sin^{-1}\sqrt{1-\xi^2}\right), \quad A = \frac{\Delta N \omega_{ref}}{N\sqrt{1-\xi^2}} \quad (4.157)$$

Thus the phase error at the kth extreme point is

$$\theta_E(t_k) = A \int_0^{t_k} f(u)\, du \quad (4.158)$$

With $\{t_k\}$ and $f(u)$ defined in (4.155) and (4.157) respectively, and after a lengthy and technical manipulation that we spare to the reader, (4.158) yields the compact form

$$\theta_E(t_k) = -\frac{\Delta N \omega_{ref}}{N \omega_n} e^{-\xi \omega_n t_k}(-1)^k, \quad 0 < \xi < 1 \quad (4.159)$$

where t_k is given in (4.155), and since $\left(e^{-\xi\pi/\sqrt{1-\xi^2}}\right)^k \le 1$ for all values of k, then

$$|\theta_E(t_k)| \le |\theta_E(t_0)| = \frac{|\Delta N|\omega_{ref}}{N\omega_n} e^{-\xi \sin^{-1}\sqrt{1-\xi^2}/\sqrt{1-\xi^2}}, \quad 0 < \xi < 1 \quad (4.160)$$

Equation (4.160) shows that the maximal absolute phase error occurs for $k = 0$, namely at the first time a plot crosses the horizontal line in Figure 4.11.

In the critically damped case, substituting (4.150) into (4.153) shows that there is only one extreme point at $t_0 = 1/\omega_n$, thus $k \equiv 0$.

Taking the limit of (4.159) when $\xi \to 1$ with $k = 0$, and noting that

$$\lim_{\xi \to 1}(\sin^{-1}\sqrt{1-\xi^2}/\sqrt{1-\xi^2}) = 1/\lim_{\xi \to 1}[\sin(\sin^{-1}\sqrt{1-\xi^2})/\sin^{-1}\sqrt{1-\xi^2}] = 1$$

$$(4.161)$$

we obtain, in the critically damped case

$$\theta_E(t_0) = -\frac{\Delta N \omega_{ref}}{N \omega_n e}, \quad t_0 = 1/\omega_n, \quad \xi = 1 \quad (4.162)$$

Figure 4.12 shows a normalized plot of $(\theta_E/\pi)/(\Delta N \omega_{ref}/N\omega_n)$ vs. $\omega_n t$. In a similar way, the extreme points of the frequency error with respect to the final value are found by equating to zero the derivative of $\Delta\omega(t)$ in (4.148) for $t > 0$.

With a short manipulation and with the help of (4.145) we obtain the extreme points $\{t_k\}$ in the form

$$\Delta\omega'(t_k) = 0 \Rightarrow t_k = \frac{k\pi + 2\sin^{-1}\sqrt{1-\xi^2}}{\omega_n\sqrt{1-\xi^2}}, \quad k \in \mathbb{N} \quad (4.163)$$

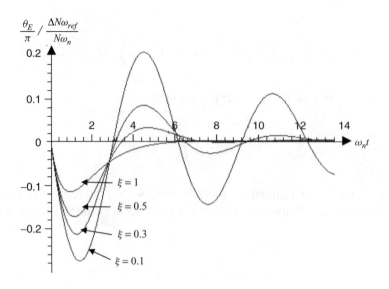

Figure 4.12   Normalized phase error transient

Substituting (4.163) into (4.148) yields

$$\frac{\Delta\omega(t_k)}{\Delta N\omega_{ref}} = 1 + e^{-\xi\omega_n t_k}(-1)^k, \ \xi < 1 \tag{4.164}$$

and since $(e^{-\xi\pi/\sqrt{1-\xi^2}})^k \leq 1$ for all values of k, then

$$|\Delta\omega(t_k) - \Delta N\omega_{ref}| \leq |\Delta\omega(t_0) - \Delta N\omega_{ref}|$$

$$= |\Delta N|\omega_{ref}e^{-2\xi\,sin^{-1}\sqrt{1-\xi^2}/\sqrt{1-\xi^2}}, 0 < \xi < 1 \tag{4.165}$$

where in the critically damped case, again with $k = 0$ and $\xi \rightarrow 1$ we get

$$\frac{\Delta\omega(t_0)}{\Delta N\omega_{ref}} = 1 + e^{-2}, \ t_0 = 2/\omega_n, \ \xi = 1 \tag{4.166}$$

4.7.3 Integer-N Lock Time Analysis

'Lock time' is a pretty loose definition depending on what is considered to be the 'lock state' of a PLL, and it is very much dependent on the actual application.

In most cases the system embedding the PLL becomes operational as soon as the VCO frequency is within some absolute error from the final state. Given the 'locked condition'

$$\left|\frac{\Delta\omega(t) - \Delta N\omega_{ref}}{\Delta N\omega_{ref}}\right| \leq \varepsilon \tag{4.167}$$

a bound to the lock time is easily inferred from (4.148) taking its absolute value and setting the requirement

$$\left| \frac{\Delta\omega(t) - \Delta N \omega_{ref}}{\Delta N \omega_{ref}} \right| \leq \frac{e^{-\xi \omega_n t_{lock}}}{\sqrt{1 - \xi^2}} = \varepsilon \qquad (4.168)$$

which leads to

$$t_{lock} \leq \frac{1}{\omega_n \xi} \ln \left(\frac{1}{\varepsilon \sqrt{1 - \xi^2}} \right) \qquad (4.169)$$

The bound (4.169) is reasonably tight if ξ is not too close to unity. An estimate at the extreme points can be obtained from (4.164). Since the absolute frequency error at the extreme points constitutes a monotonically decreasing sequence, then the requirement

$$\left| \frac{\Delta\omega(t_k) - \Delta N \omega_{ref}}{\Delta N \omega_{ref}} \right| = e^{-\xi \omega_n t_k} \leq \varepsilon, \ \xi < 1 \qquad (4.170)$$

leads to the lock time bound

$$t_{lock} \leq \min_{k \in \mathbb{N}} \{t_k\} | t_k \geq \frac{1}{\omega_n \xi} \ln \left(\frac{1}{\varepsilon} \right), \ \xi < 1 \qquad (4.171)$$

Using (4.163) and (4.171) k is the smallest integer satisfying

$$k \geq \frac{1}{\pi} \left[\frac{\sqrt{1 - \xi^2}}{\xi} \ln \left(\frac{1}{\varepsilon} \right) - 2 \sin^{-1} \sqrt{1 - \xi^2} \right] \qquad (4.172)$$

Since (4.132) implies

$$\omega_n \xi = \frac{1}{2}\alpha, \ \alpha = \frac{K_\phi K_V K_L}{N}, \ \xi < 1 \qquad (4.173)$$

as long as $\xi < 1$, (4.132) implies that

$$t_{lock} \leq \min_{k \in \mathbb{N}} \{t_k\} | t_k \geq \frac{2N}{K_\phi K_V K_L} \ln \left(\frac{1}{\varepsilon} \right), \ \xi < 1 \qquad (4.174)$$

Equation (4.174) shows that we may reduce the lock time by just reducing N without increasing the VCO sensitivity K_V (which would increase the reference spur level and make the VCO more prone to parasitic frequency modulation if a small interference reaches the steering line). The extension of the above computations to the fractional-N PLL will show that the beneficial effect of a small divider is indeed exploited there.

Example: Consider an integer-N synthesizer working at 820.025 MHz with a reference frequency of 25 kHz (and consequently with a frequency resolution of

25 kHz). Let us compute the time required to jump 13.9 MHz to the frequency 806.125 MHz, within a 500 Hz error.

- The required division before jump is $N = 820025/25 = 32801$.
- The required division after jump is $N = 806125/25 = 32245$.

Therefore

$$\Delta N = 32801 - 32245 = 556$$

$$\varepsilon = 0.5/13900 \approx 3.6 \times 10^{-5}$$

Assume

$$K\phi = 0.1 \text{ mA/rad}$$

$$K_L = 2k\Omega$$

$$K_V = 8 \text{ MHz/V}$$

Then

$$\alpha = 16 \times 10^5/32245 \approx 49.6$$

Choosing $\omega_L = \alpha$, (4.132) leads to

$$\xi = 0.5, \quad \omega_n = \omega_L = \alpha$$

In order to filter out the spurious components due to the phase detector pulses, it is customary to add a pre-integration capacitor that introduces the parasitic pole ω_p described in (4.128) and (4.129). As pointed out before, the additional pole must yield a loop filter bandwidth much smaller than the reference frequency, but much higher than ω_n.

If we choose $\omega_p = 10\alpha \approx 500$ so it may be neglected for the purpose of lock-up time computations, the loop filter bandwidth is about 300 times lower than the reference frequency, which ensures about 25 dB attenuation of reference spurs. The influence of the parasitic pole on PLL stability is analyzed in Chapter 6.

Using (4.173) we get an estimate of the lock time from (4.171) as

$$t_{lock} \approx -ln(3.6 \times 10^{-5})/(49.6/2) \approx 0.41s$$

Now using (4.169) we can bound

$$t_{lock} \leq -ln(3.6 \times 10^{-5} \times 0.71)/(49.6/2) \approx 0.43s$$

As it can be expected by comparing (4.169) and (4.171), both values are very close because the difference is within the argument of the logarithm. We conclude that to get a first estimate of the lock time, we do not need to know a priori the values of ω_n and ξ, but we only need to compute their product $\omega_n\xi = \alpha/2$ using (4.173).

If we had to jump just 25 kHz (one channel), we would have $\Delta N = 1$ and therefore $\epsilon = 0.5/25 = 0.02$, which, leads to

$$t_{lock} \approx -\ln(0.02)/(49.6/2) \approx 0.16\text{s}$$

The figures given in this example are representative of the long lock times required by integer-N synthesizers. Such long times, although reasonable for certain systems, are incompatible with the requirements of fast-hopping equipment used in many modern standards.

4.7.4 Phase-Frequency Detector

In view of its importance in understanding the behavior of a digital synthesizer, we outline a short description of the operation of the phase-frequency detector. One of its many possible implementations is shown in Figure 4.13, where the input signals are either at logic '1' or logic '0' level, and:

- Both flip-flop are of the D-type, namely upon a rising edge at the Clk input, their Q output becomes a logic '1', and only upon applying a logic '1' at the *Reset* input, does the Q output becomes a logic '0'.
- $F_R(t)$ is the 'square-wave' reference oscillator signal.
- $F_V(t)$ is the 'square-wave' divided VCO signal
- The upper CMOS device acts as a fixed current *source* supplying I_{dc} when Q_{UP} is '1', and is an open circuit when Q_{UP} is '0'.
- The lower CMOS device acts as a fixed current *sink* driving I_{dc} when Q_{DN} is '1', and is an open circuit when Q_{DN} is '0'.
- The loop filter is the low-pass filter of Figure 4.10.

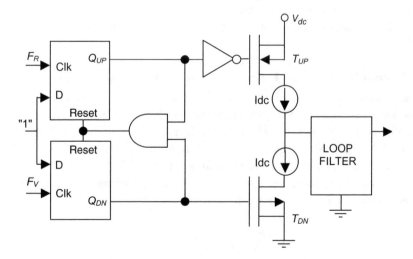

Figure 4.13   Phase-frequency detector

Phase mode (see Figure 4.13). Assume that at the beginning of the process, at $t < 0$, F_R and F_V have exactly the same frequency and the same phase. At the time $t = 0$, the programmable counter is changed, thus f_R changes. Assume, without loss of generality, that the resulting frequency f_V of F_V is lower than the fixed frequency f_R of F_R. The phase mode of operation occurs as long as the input frequencies are close enough so that $0.5 < f_R/f_V < 2$.

Assume that a '*Reset*' has just occurred, then both Q_{UP} and Q_{DN} are at logic '0'. In this state, both T_{UP} and T_{DN} are open. Assume also that the voltage output of the loop filter has stabilized at some value between zero and Vdc.

Since $f_R > f_V$, the rising edge of F_R occurs first, then Q_{UP} is set and T_{UP} goes into the conduction. When the rising edge of F_V occurs, Q_{DN} is set, the AND gate is activated and both flip-flop are reset almost instantly. Thus, T_{UP} has been active for the time (phase) difference between the rising edges of F_R and the rising edge of F_V, and T_{DN} has been in active state for the negligible time required for the AND gate to react and reset both Q_{UP} and Q_{DN} to '0'. It follows that, as long as $f_R > f_V$, the input to the loop filter is a train of source current pulses tending to push the output voltage up.

It is left to the reader to show that in the case $f_R < f_V$, we end up with a train of sink pulses tending to push the output voltage down.

Frequency mode: the frequency mode of operation occurs when the frequencies are far enough so that $f_R/f_V < 0.5$ or $f_R/f_V > 2$.

Assume the same starting conditions as before, and assume without loss of generality that $f_R/f_V > 2$. In this case, more than one period of F_R will elapse from a rising edge of F_R and until a rising edge of F_V causes the reset to be activated. Since the time difference between the rising edge of F_V to the rising edge of F_R is shorter than one period of F_R, it can be shown that, on average, the current pulse at the input of the loop filter will be in source mode for a longer time than in sink mode. As a result, the average current will be of the source type, tending to push the output voltage up. Again, the opposite case is left to the reader to show.

4.8 Extension of PLL Fundamentals to Fractional-N

We show now that in the fractional-N case, all the equations developed in section 4.7.2 and 4.7.3 hold approximately with N replaced by $N + M/D$ and ΔN replaced by $\Delta N + \Delta M/D$, with N, M and D as defined in Equation (4.6), and with ΔN, ΔM being the changes occurring in the values of N and M at the instant $t = 0$, just as described in section 4.7.1 and 4.7.2.

Since in the fractional-N setting the value of the divider is a function of time, we may rewrite (4.111) in the form

$$\theta_{ref}(t) - \theta_{ref}(0) = \int_0^t \omega_{ref} \, d\xi = \omega_{ref} t$$

$$\theta_{out}(t) - \theta_{out}(0) = \int_0^t \frac{\omega_0 + \Delta\omega(\xi)}{N(\xi)} d\xi \tag{4.175}$$

where

$$N(t) = N + \Delta N + \Delta_n[u(t - nT_R) - u(t - (n+1)T_R)]$$

$$M + \Delta M = \sum_{n=1}^D \Delta_n < D \tag{4.176}$$

As before, assume steady-state lock for $t < 0$ and phase coherence at $t = 0$ and denote by $\theta_E(t) = \theta_{ref}(t) - \theta_{out}(t)$ the phase error out of the phase detector.

With a reference clock period T_R, assuming that $\theta_E(t)$ is much slower than the sampling rate $f_R = 1/T_R$, we can use the first-order Taylor expansion

$$\theta_E(t + (n+1)T_R) \approx \theta_E(t + nT_R) + \theta'_E(t + nT_R) \cdot T_R \tag{4.177}$$

Then, with the help of (4.175), (4.176) and (4.177), we get

$$\theta'_E(t + nT_R) \cdot T_R \approx \theta_E(t + (n+1)T_R) - \theta_E(t + nT_R)$$

$$= \omega_{ref} T_R - \int_{t+nT_R}^{t+(n+1)T_R} \frac{\omega_0 + \Delta\omega(\xi)}{N + \Delta N + \Delta_n} d\xi$$

$$\approx \omega_{ref} T_R - \frac{\omega_0 + \Delta\omega(\lambda)}{N + \Delta N + \Delta_n} T_R \tag{4.178}$$

where in (4.178)

$$\lambda \in (t + nT_R, t + (n+1)T_R) \tag{4.179}$$

and we used the mean-value theorem for integrals

$$\int_a^b f(x)\, dx = (b-a)f(\lambda), \quad \lambda \in (a, b) \tag{4.180}$$

Assuming that $\theta'_E(t)$ and $\Delta\omega(t)$ may be considered nearly constant over a counting pattern period DT_R (which will be justified if the loop filter has a bandwidth of about $1/DT_R$), namely

$$\theta'_E(t + nT_R) \approx \theta'_E(t), \quad \Delta\omega(\lambda) \approx \Delta\omega(t), \quad n = 1, 2, \ldots, D \tag{4.181}$$

Then, multiplying (4.178) by $N + \Delta N + \Delta_n$, summing up for $n = 1, 2, \ldots, D$, and since

$$\sum_{n=1}^D (N + \Delta N + \Delta_n) = D(N + \Delta N) + \sum_{n=1}^D \Delta_n = D(N + \Delta N) + M + \Delta M$$

$$\tag{4.182}$$

we can manipulate (4.178) to the form

$$\theta'_E(t) \approx \omega_{ref} - \frac{\omega_0 + \Delta\omega(t)}{\left(N + \dfrac{M}{D}\right) + \Delta N + \dfrac{\Delta M}{D}} \tag{4.183}$$

Assuming that

$$\left| \Delta N + \frac{\Delta M}{D} \right| \ll N + \frac{M}{D} \tag{4.184}$$

we may rewrite (4.183) as

$$\theta'_E(t) \approx \omega_{ref} - \frac{\omega_0}{N + \dfrac{M}{D}}\left(1 - \frac{\Delta N + \dfrac{\Delta M}{D}}{N + \dfrac{M}{D}}\right) - \frac{\Delta\omega(t)}{\left(N + \dfrac{M}{D}\right)}\left(1 - \frac{\Delta N + \dfrac{\Delta M}{D}}{N + \dfrac{M}{D}}\right) \tag{4.185}$$

and noting that

$$\omega_{ref} = \frac{\omega_0}{N + \dfrac{M}{D}}, \quad \frac{\Delta\omega(t)}{\left(N + \dfrac{M}{D}\right)}\left(1 - \frac{\Delta N + \dfrac{\Delta M}{D}}{N + \dfrac{M}{D}}\right) \approx \frac{\Delta\omega(t)}{\left(N + \dfrac{M}{D}\right)} \tag{4.186}$$

we finally obtain from (4.185)

$$\theta'_E(t) \approx \frac{\Delta N + \dfrac{\Delta M}{D}}{N + \dfrac{M}{D}}\omega_{ref} - \frac{\Delta\omega(t)}{\left(N + \dfrac{M}{D}\right)} \tag{4.187}$$

Integrating (4.187) on $[0, t]$ we finally obtain (4.116) with N replaced by $N + M/D$ and ΔN replaced by $\Delta N + \Delta M/D$, namely

$$\theta_E(t) \approx \frac{\Delta N + \dfrac{\Delta M}{D}}{N + \dfrac{M}{D}}\omega_{ref}t \cdot u(t) - \int_0^t \frac{\Delta\omega(\xi)}{N + \dfrac{M}{D}}d\xi \tag{4.188}$$

Since all the results for the integer-N theory follow from (4.116), the same results apply approximately for the fractional-N case with the substitution

$$N \to N + \frac{M}{D}, \Delta N \to \Delta N + \frac{\Delta M}{D} \tag{4.189}$$

Example: Consider the same example of section 4.7.3, but using a fractional-N synthesizer with a reference frequency of 2.1 MHz. As before, let us compute the time required to jump 13.9 MHz from 820.025 MHz to 806.125 MHz within 500 Hz error. Since the synthesizer must have a 25 kHz resolution, then:

$$D = 2.1 \times 10^6/25 \times 10^3 = 84$$

with square brackets indicating an integer value we have

- Before jump:

$$N = [820.025/2.1] = 390$$

$$M = (820.025 - N \times 2.1)D/2.1 = 41$$

- After jump:

$$N = [806.125/2.1] = 383$$

$$M = (806.125 - N \times 2.1)D/2.1 = 73$$

Therefore $\Delta N = 7$, $\Delta M = -32$ and $\varepsilon = 0.5/13900 \approx 3.6 \times 10^{-5}$. Assume, as before

$$K\phi = 0.1 \text{ mA/rad}$$

$$K_L = 2\text{k}\Omega$$

$$K_V = 8 \text{ MHz/V}$$

Then

$$\alpha = 16 \times 10^5/383 \approx 4177$$

Again choosing $\omega_L = \alpha$, (4.132) leads to

$$\xi = 0.5, \quad \omega_n = \omega_L = \alpha$$

and the estimate of the lock time from (4.171) is

$$t_{lock} \approx -ln(3.6 \times 10^{-5})/(4177/2) \approx 0.005\text{s}$$

As expected, the lock time is about 100 times faster than in the integer-N implementation, which is just the ratio of the values of N in the two cases.

Choosing the parasitic pole $\omega_p = 10$, $\alpha \approx 42,000$ again yields a loop filter bandwidth about 300 times lower than the reference frequency, and also satisfies the requirement (4.181). However, it must be remembered that, as pointed out before, the attenuation of the reference spurs in the fractional-N case is very much related to the counting pattern (which determines the reference noise shape), and does not depend only on the loop filter bandwidth.

4.9 Measurement of Synthesizers

The main concern in synthesizer applications is usually related to either lock-up time or spectral purity. Nevertheless, since ultimately we are dealing with an

RF carrier, there is much in common between the measurement of synthesizers and the measurement of oscillators, and, whenever applicable, we refer to the corresponding setups and methods described in Chapter 5.

4.9.1 Lock Time

As pointed out before, the definition of lock time is not unique, and is very much dependent on the actual application.

In narrowband transceivers, such as analog FM radios with channel spacing of 12.5 kHz and channel bandwidth of 9 kHz, the synthesizer is considered in lock state when the VCO frequency is within less than 1 kHz from the final frequency. In contrast, for wideband transceivers such as in WLAN, the system may already become operational when the synthesizer is still many kHz away from the final frequency.

Denote by f_0 the steady-state frequency at the instant $t = 0$, just at the start of a synthesizer 'jump', by f_∞ the steady-state frequency at time $t \to \infty$, and by f, the instantaneous frequency for $t \geq 0$.

In to avoid ambiguity, we must define:

- The frequency jump
$$f_{jump} = f_\infty - f_0$$

- The lock-state frequency deviation Δf_{lock}
$$|f - f_\infty| < \Delta f_{lock}, t \to \infty$$

- The lock time t_{lock}
$$t > t_{lock} \to |f - f_\infty| < \Delta f_{lock}$$

So, given the frequency jump and the allowed frequency deviation, the lock time is the instant after which $|f - f_\infty|$ no longer exceeds Δf_{lock}.

Several methods can be used to measure t_{lock}. The simplest and most accurate is the 'mixer method', whose setup is described in Figure 4.14.

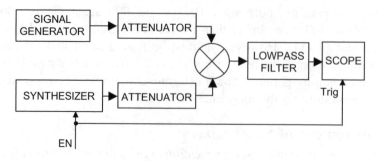

Figure 4.14 The mixer method setup

The mixer is of the passive double-balanced type. The frequencies to be measured by the oscilloscope are in the baseband range, and are typically very low (of the order of few kHz). The low-pass filter may be of any kind (even a simple RC filter will do). The attenuators (≥ 10 dB each) are an absolute requirement. This is because the passive mixer needs a relatively high drive level (larger than 0 dBm), and without the attenuators, mutual injection locking between the signal generator and the VCO may occur (Chapter 6) leading to erroneous measurement results.

The low-pass filter has the purpose of filtering out the high frequency portion of the mixer output, and usually has a corner frequency of few tens of kHz at most. Thus, we will start to see substantial output on the oscilloscope only when the synthesizer frequency becomes close to the RF generator.

It should be noted that, due to tolerances in the reference oscillators, even under stable lock conditions there will be always some frequency difference between the signal generator and the synthesizer.

We recommend setting the signal generator at some small offset from the final frequency of the synthesizer, say, at $f_\infty + 500$ Hz. This will ease the reading by masking the low-frequency fluctuations of the generator. With this setting, when we sees a 500 Hz signal on the scope, the synthesizer has reached f_∞. The picture on the oscilloscope will look as shown in Figure 4.15.

The measurement proceeds as follows:

- The signal generator is set to $f_\infty + 500$ Hz, and the synthesizer is first set at the initial frequency f_0. Since the frequency difference $|f_\infty - f_0|$ is much larger than the bandwidth of the low-pass filter, at this point no signal is seen on the oscilloscope.
- The *EN* (enable) signal is set, which allows the loading of the new frequency f_∞ to the synthesizer, and also triggers the oscilloscope.

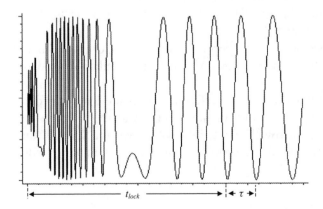

Figure 4.15   The mixer method oscilloscope picture

- The lock time t_{lock} is the time elapsed from the instant f_∞ was loaded into the synthesizer, to the instant at which $1/\tau$ exceeded the value $\Delta f_{lock} + 500$ Hz for the last time.

4.9.2 Frequency Accuracy and Stability

By 'frequency accuracy and stability' we mean the frequency deviation of the synthesizer from its nominal working frequency, under temperature and voltage variations.

For temperature and voltage, the test is carried out by putting the synthesizer into a heat-cool oven. First the output frequency is measured at 'room' temperature (nominally $+25\,°C$) and at nominal, maximal and minimal DC supply voltages. Next, the measurement is repeated at extreme temperatures, usually $-30\,°C$ and $+60\,°C$. Finally, the frequency accuracy and stability of the synthesizer is the largest absolute deviation from the nominal frequency, measured under all temperature and voltage variations, and is usually given in PPM (parts per million).

4.9.3 Reference Spurs

By 'reference spurs' we refer to the spurs generated in the VCO output by the phase detector pulses. As we saw before, these pulses are attenuated by the loop filter and by the pre-integration capacitor, but their residuals still reach the VCO steering line, producing narrowband FM modulation.

In integer-N synthesizers, this parasitic FM modulation shows up in the form of low-level carriers, shifted from both sides of the center carrier by multiples of the reference frequency, from which the name 'reference spurs' is derived. In the fractional-N setting, as seen before, the spectral picture of the reference spurs is strongly dependent on the counting pattern.

The measurement of reference spurs can be done either by directly observing the neighborhoods of the carrier using a high-frequency spectrum analyzer, or with a procedure identical to the one used for measuring the SSB noise of an oscillator in section 5.7.2.

The reference spurs will 'pop up' from the VCO noise floor in a way similar to the intentional parasitic narrowband FM modulation used for comparison in Chapter 5.

It should be noted that if the reference spurs are low, in order to be able to see them the IF filter bandwidth of the spectrum analyzer has to be set narrow. On the one hand, a narrow IF bandwidth will lower the level of the SSB noise thus 'discovering' the reference spurs, but on the other hand, it will make the sweep of spectrum analyzer sluggish.

5

Oscillators

Oscillator design is always a critical issue for RF engineers. At first sight, building a circuit that oscillates at some RF frequency covering a defined range and delivering a specified power output is not a difficult task. The real challenge, however, is to achieve low close-in phase noise required for receiver selectivity (section 2.3), a low noise floor essential for obtaining blocking immunity (section 2.8), and transmitter spectral purity, while avoiding parasitic effects such as remodulation, load pulling, injection pulling, microphonics etc. (Chapter 6), and while attaining the lowest possible current consumption. Needless to say, the above requirements are all conflicting, and each one is achieved at the expense of the others.

Parasitic effects of the kind specified above, unless taken in account in the very first design stage, cannot be corrected *a posteriori*, and often show up during equipment production or shipment and can lead to last-minute disasters.

On the one hand, from the standpoint of current drain, operating band, size and cost, the VCO performance requirements are so diversified and application dependent, that picking up off-the-shelf modules is rarely an option. On the other hand, the theory and know-how of low-noise VCO design are tough matters that seldom can be found in an organized, comprehensive, and easy-to-understand format.

Due to the diversified nature of the subject, we have found it essential to include in this chapter quite a bit of theoretical material. However, we have made an effort to reduce it to a minimum, and, whenever possible, we try to keep the pace and to ease the global understanding by making use of the applicable results without proof, and we defer further insight to the bibliography.

For the sake of clarity, we provide a very detailed VCO design example using a low-power BJT (bipolar junction transistor). In the framework of this example we also develop the general equations required for the design of bipolar oscillators.

CMOS-gate oscillators follow a similar procedure, as will be shown later, while treating the design of inverter-gate driven crystal oscillators. More insight may be found in Rohde (1997).

Wireless Transceiver Design   Ariel Luzzatto and Gadi Shirazi
© 2007 John Wiley & Sons, Ltd

5.1 Low-Power Self-Limiting Oscillators

All the practical oscillators we deal with, whether tunable or not, consist essentially of the elements shown in Figure 5.1, namely:

(I) A voltage-controlled nonlinear current source driven by a controlling input voltage $v_i(t) = V \cos(\omega_i t)$, where V can be assumed constant over time as compared to the angular frequency ω_i.

Denoting by $V_i(\omega)$ the Fourier transform of the signal $v_i(t)$ applied to the input (which we refer to as 'the limiting port'), the Fourier transform of the output current $i_0(t)$ is given by $I_0(\omega) = G(V)V_i(\omega)$, where $G(V)$ is a nonlinear function of the peak amplitude V, monotonically decreasing in absolute value as V increases.

(II) A trans-impedance (current-in, voltage-out) resonant linear feedback network with transfer function $Z(\omega)$, that, for $\omega > 0$, has the form

$$Z(\omega) = R/(1 + j2Q\Delta\omega/\omega_0), \quad \omega > 0 \tag{5.1}$$

where R is a real constant, ω_0 is referred to as 'the resonant frequency' and is often a controllable (tuned) parameter. $\Delta\omega = \omega - \omega_0$ is the frequency deviation from resonance, and Q, known as 'the quality factor at resonance', is a measure of the resistive loss of the network, and may or may not include the effect of resistive losses caused by external circuits interacting with it. In general, the larger the value of Q, the better the oscillator performance.

When only the losses of the resonant network itself are taken into account then we refer to the quality factor as to 'the unloaded Q' and we designate it with the symbol Q_U.

The unloaded Q is a measure of the losses that are 'unavoidable', in the sense that they are an inherent part of the technology used for building the resonant network itself, and we cannot improve them by means of a better circuit design.

When we incorporate within the resonant circuit all the losses introduced into the circuit due to passive and active components that are not a part of the network itself, but interact directly with it, then we refer to the quality factor as to

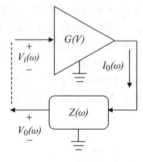

Figure 5.1   Self-limiting oscillator architecture

'the loaded Q' and we designate it with the symbol Q_L. The loaded Q can be controlled, to some extent, by careful circuit design. Always $Q_U \geq Q_L$.

If we connect the output of the current source to the input of the feedback network, the voltage $V_0(\omega)$ at the output of the feedback network can be written in the form

$$V_0(\omega) = Z(\omega)I_0(\omega) = Z(\omega)G(V)V_i(\omega) \qquad (5.2)$$

Figure 5.1 illustrates the outcome of Equation (5.2).

If $G(V)$ can adjust itself so that in (5.2) $V_0(\omega) = V_i(\omega)$ for some value of ω and V, then the output of the feedback network can be connected to the input of the current source, and the output current $I_0(\omega)$ will continue to exist without the need of an external input signal $V_i(\omega)$, which, in other words, means that oscillations occur. Thus, it follows from (5.2) that an oscillation state must satisfy the complex-valued nonlinear equation

$$Z(\omega)G(V) = 1 \qquad (5.3)$$

known as 'the Barkhausen criterion'.

The requirement $Im[Z(\omega)G(V)] = 0$ determines the oscillation frequency ω_0, while the requirement $Re[Z(\omega)G(V)] = 1$ determines the amplitude V of the oscillations. In other words, (5.3) states that the frequency and the amplitude of the oscillations self-stabilize so that the loop gain is real and equal unity.

The most practical feedback network $Z(\omega)$, which we employ in all our low-power designs, uses a 'π-topology'. Figure 5.2 shows the oscillator of Figure 5.1 with the π-topology network.

Figures 5.3 and 5.4 show that different oscillator topologies, apart from the choice of the grounding point, are equivalent to the π-topology oscillator of Figure 5.2. Therefore, from now on, we always reduce oscillators to the form of Figure 5.2, which is the most convenient for computations.

For the sake of early insight, we anticipate the following:

1. The nonlinear transfer function $G(V)$ determines the operating point of the oscillator.
2. The critical value to design is the peak oscillating voltage at the limiting port of the nonlinear current source.

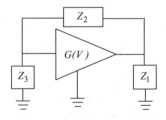

Figure 5.2 π-topology oscillator

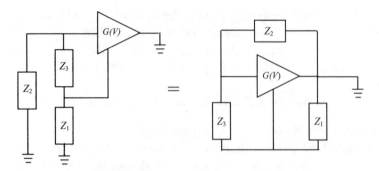

Figure 5.3 Colpitts π-topology equivalence

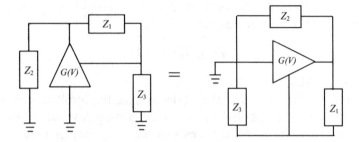

Figure 5.4 'Common input' π-topology equivalence

3. The oscillating voltage amplitude at the limiting port determines the far-out phase noise of the oscillator (the noise floor).
4. The quality factor of the transfer function $Z(\omega)$ including input and output loading (the 'loaded Q'), determines the close-in phase noise.

5.2 Feedback Network Design

We represent the feedback network, whether tuned or not, and whether implemented using lumped components, transmission lines, piezoelectric crystals, distributed arrangements or any other means, by the equivalent lumped model of Figure 5.5, where $X_R(\omega)$ is an arbitrary inductive reactance which, we assume, can be approximated in the range of interest, by a first-order Taylor expansion as shown in (5.4), where $L(\omega)$ is a real-valued function of ω

$$X_R(\omega) = jL(\omega), \ L(\omega) \approx L(\omega_0) + L'(\omega_0)(\omega - \omega_0)$$

$$L(\omega_0) = \frac{1}{\omega_0}\left(\frac{1}{C_I} + \frac{1}{C_V}\right), \ L'(\omega) \equiv \frac{dL(\omega)}{d\omega}, \ |\omega - \omega_0|/\omega_0 \ll 1 \quad (5.4)$$

In Figure 5.5, the resistor r appearing in series to the reactance, represents all the losses due to the network, including the power delivered to load. In the detailed design examples we will explain how to incorporate these losses within r.

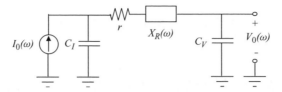

Figure 5.5   Lumped feedback network (loaded)

Denoting by $Z(\omega) = V_0(\omega)/I_0(\omega)$ the trans-impedance of the network, and by using the value of $L(\omega_0)$ in (5.4), it is straightforward to show that:

1. The circuit is resonant at $\omega = \omega_0$. In other words $Z(\omega_0)$ is a real number.
2. The transfer function at resonance is $Z(\omega_0) = -1/(\omega_0^2 C_I C_V r)$ for *any* $X_R(\omega)$ that satisfies (5.4).

Setting $\omega = \omega_0[1 + (\omega - \omega_0)/\omega_0]$ we may write

$$Z(\omega) = \frac{V_0(\omega)}{I_0(\omega)} \approx -\frac{1}{\omega_0^2 C_I C_V r} \frac{1}{1 + j\frac{1}{r}[L(\omega_0) + \omega_0 L'(\omega_0)]\frac{\omega - \omega_0}{\omega_0}} \qquad (5.5)$$

We prove (5.5) as follows: setting

$$Z_I = 1/j\omega C_I, \quad Z_V = 1/j\omega C_V, \quad \Delta\omega = \omega - \omega_0$$

and using (5.4) we may write

$$\frac{V_0(\omega)}{I_0(\omega)} = \frac{Z_I(X_R + r + Z_V)}{Z_I + X_R + r + Z_V} \frac{Z_V}{X_R + r + Z_V} = \frac{Z_I Z_V}{Z_I + X_R + r + Z_V}$$

$$= -\frac{1}{\omega^2 C_I C_V} \frac{1}{r - j\frac{1}{\omega}\left(\frac{1}{C_I} + \frac{1}{C_V}\right) + jL(\omega_0) + jL'(\omega_0)\Delta\omega}$$

In the range of interest, we assume $\left|\dfrac{\Delta\omega}{\omega_0}\right| \ll 1$, then

$$\omega = \omega_0\left(1 + \frac{\Delta\omega}{\omega_0}\right) \Rightarrow \frac{1}{\omega} = \frac{1}{\omega_0\left(1 + \dfrac{\Delta\omega}{\omega_0}\right)} \approx \frac{1}{\omega_0}\left(1 - \frac{\Delta\omega}{\omega_0}\right)$$

Substituting for $1/\omega$, the previous expression yields

$$\frac{V_0(\omega)}{I_0(\omega)} \approx -\frac{1}{\omega_0^2 C_I C_V}$$

$$\times \frac{1}{r - j\frac{1}{\omega_0}\left(\frac{1}{C_I} + \frac{1}{C_V}\right)\left(1 - \frac{\Delta\omega}{\omega_0}\right) + jL(\omega_0) + j\omega_0 L'(\omega_0)\frac{\Delta\omega}{\omega_0}}$$

By (5.4) $\dfrac{1}{\omega_0}\left(\dfrac{1}{C_I}+\dfrac{1}{C_V}\right)=L(\omega_0)$, then the last equation becomes

$$\frac{V_0(\omega)}{I_0(\omega)}\approx-\frac{1}{\omega_0^2 C_I C_V}\frac{1}{r+jL(\omega_0)\dfrac{\Delta\omega}{\omega_0}+j\omega_0 L'(\omega_0)\dfrac{\Delta\omega}{\omega_0}}$$

which yields directly (5.5).

A trivial example is when $L(\omega)=\omega L$, then $X_R(\omega)=j\omega L$ is the reactance due to a fixed inductor L. In this case we obtain from (5.5) the well-known approximation

$$Z(\omega)\approx-\frac{1}{\omega_0^2 C_I C_V r}\frac{1}{1+j2Q_L\dfrac{\omega-\omega_0}{\omega_0}}$$

$$Q_L=\frac{\omega_0 L}{r},\ \omega_0^2=\frac{1}{L}\left(\frac{1}{C_V}+\frac{1}{C_I}\right)\tag{5.6}$$

A less trivial, but very important case, is when $X_R(\omega)$ is a series resonant LC circuit as shown in Figure 5.6. Here, $X_R(\omega)=j(\omega L-1/\omega C)$ is the impedance of the LC circuit. Using the constraints given in the bottom equation of (5.4) for $L(\omega)$ we get

$$L(\omega)=\omega L-\frac{1}{\omega C},\ L(\omega_0)+\omega_0 L'(\omega_0)=2\omega_0 L$$

$$L(\omega_0)=\frac{1}{\omega_0}\left(\frac{1}{C_I}+\frac{1}{C_V}\right)\Rightarrow\omega_0^2=\frac{1}{L}\left(\frac{1}{C}+\frac{1}{C_V}+\frac{1}{C_I}\right)\tag{5.7}$$

Then, with the help of (5.7), (5.5) yields

$$Z(\omega)\approx-\frac{1}{\omega_0^2 C_I C_V r}\frac{1}{1+j2Q_L\dfrac{\omega-\omega_0}{\omega_0}}$$

$$Q_L=\frac{\omega_0 L}{r},\ \omega_0^2=\frac{1}{L}\left(\frac{1}{C}+\frac{1}{C_V}+\frac{1}{C_I}\right)\tag{5.8}$$

which, apart from the resonating frequency, is identical to (5.6).

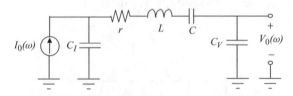

Figure 5.6   Using a serial resonant circuit for $X_R(\omega)$

In spite of the apparent similarity, we will see later that the last configuration is the most useful in oscillator design.

Since $Z(\omega_0) = -1/(\omega_0^2 C_I C_V r)$ regardless of $X_R(\omega_0)$, it follows from (5.3) that, if there exists a value of V that satisfies the nonlinear equation

$$G(V) = -\omega_0^2 C_I C_V r \qquad (5.9)$$

oscillations will occur and self-stabilize at that value. Note that the negative sign in equation (5.9) implies that we must use a phase-inverting active device. As pointed out in section 2.1, $|G(V)|$ is a monotonically decreasing function. Thus, whenever V is smaller than the self-limiting value, the amplitude of the oscillations will continue to grow, which assures oscillator start-up following the presence of any small disturbance such as thermally generated noise.

5.3 Noisy Oscillator – Leeson's Equation

Oscillator phase noise is of fundamental importance in most RF designs, therefore, in spite of the pretty complex mathematical treatment, we find it worthwhile to show an analysis that leads to the Leeson's equation describing the noise characteristics of an oscillator.

To prove the equation, it is interesting to use a model based on stochastic processes, in lieu of more conventional semi-deterministic strategies often employed.

Whenever required along the way, we refresh 'ad hoc' the concepts employed for the analysis. Nevertheless, readers without a background on stochastic processes should refer to the bibliography for a deeper understanding.

We will prove here the Leeson's equation (5.10):

$$L(\omega - \omega_0) = \log_{10}\left\{\frac{1}{2}\frac{kTF}{P_S}\left[1 + 1/\left(2Q_L\frac{\omega - \omega_0}{\omega_0}\right)^2\right]\right\} \quad [dBc/Hz] \qquad (5.10)$$

where
$\omega_0 > 0$ is the oscillator center frequency, and $0 < \omega \neq \omega_0$.
$L(\omega - \omega_0)$ denotes the phase noise in [dBc/Hz] at frequency ω.
P_S is the oscillator power at the limiting port.
Q_L denotes the loaded Q of the feedback network as in section 5.1.
k is the Boltzmann constant in MKS units.
F is the (numeric) device noise figure ($NF = 10\log_{10} F$).
T is the temperature in degrees Kelvin.
kTF is the noise power density into the input of the limiting port.

Denoting by $V_{N,rms}$ the equivalent rms noise voltage density in $[V/(Hz)^{1/2}]$, and by $V_{S,rms}$ the equivalent rms signal voltage seen at the input of the limiting port, then we may set

$$\frac{kTF}{P_S} = \frac{V_{N,rms}^2}{V_{S,rms}^2} \qquad (5.11)$$

From Equation (5.10), the following important considerations arise:

- When $|\omega - \omega_0| \to \infty$, namely far away from the oscillator frequency, the 'noise floor' depends on the signal power at the limiting port.
- The larger the oscillation amplitude, the better the noise floor. Of course this cannot be achieved for free. When designing the different examples we will show how to optimize the oscillation amplitude.
- When $|\omega - \omega_0| \to 0$, namely near the oscillator frequency, the 'close-in' noise depends mainly on the loaded Q of the feedback network. Again, in the design examples we will show the considerations for optimizing this parameter for different applications.
- The close-in phase noise decreases by about 6 dB/octave away from the center frequency.

When looking very close to the center frequency, say for $|\omega - \omega_0| < \omega_f$, where ω_f depends on device technology, and is referred to as 'the flicker corner', the noise figure deteriorates, and we should substitute in (5.10)

$$F \longrightarrow F\left(1 + \frac{\omega_f}{|\omega - \omega_0|}\right) \tag{5.12}$$

Due to the form of (5.12), the noise below the flicker corner is often referred to as the '1/f noise'.

The flicker corner in bipolar transistors is around 1 kHz, thus its effect is negligible in most applications, while in CMOS devices a flicker corner of several hundreds of kHz is a common figure, and it must be accounted for.

Let us see how we get (5.10). In Figure 5.7, G is the device gain at oscillations, $h(t)$ is the impulse response of the feedback network and $H(\omega)$ is its Fourier transform. Since at oscillation frequency ω_0 the Barkhausen criterion must hold, assuming that $H(\omega_0) = 1$ we take $G \equiv 1$.

We make the following assumptions:

- $n(t)$ is the thermal noise signal, and belongs to a white random process N with power $N_0/2$ watt/Hz.

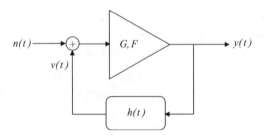

Figure 5.7   Model of a noisy oscillator ($G \equiv 1$)

- $N_0 = kTF$, where F denotes the noise figure of the active device. N_0 is the equivalent (one-sided) noise power density when the effect of the noise figure is referred to the input and the device is considered noiseless.
- $h(t)$ is the (real-valued) impulse response of the feedback (band-pass) network.
- $v(t)$ is the RF feedback signal at the (input) limiting port. Since $v(t)$ includes the effects of the random process $n(t)$ and of the band-pass filter $h(t)$, it belongs to a (band-pass) band-limited random process V.

Before we start, let us recall some definitions related to random processes (Schwartz, 1990):

- A random process X consists of a collection of many possible functions (sample functions) $x(t)$ that *might* have occurred.
- Let $x(t)$ be a sample function from a stationary process X, then at a given time t_0, the sample value $x(t_0)$ is a random variable (RV).
- $E[x(t_0)]$ denotes the expected value (the statistical average) of $x(t_0)$.
- $R_{XX}(\tau) \equiv E[x(t)x(t+\tau)]$ denotes the autocorrelation of the RV $x(t)$.
- $R_{XY}(\tau) \equiv E[x(t)y(t+\tau)]$ is the cross-correlation between the RVs $x(t)$ and $y(t)$.
- If X is white, then $R_{XX}(\tau) = (N_0/2)\delta(\tau)$, where $\delta(\tau)$ denotes the Dirac distribution.
- The Fourier transforms of $R_{XX}(\tau)$ and $R_{XY}(\tau)$ denote respectively the spectral density of $x(t)$ and the cross-spectral density of $x(t)$ and $y(t)$, with the notation

$$\hat{R}_{AB}(\omega) = Fourier\{R_{AB}(\tau)\} \quad [watt/Hz]$$

If a random process X goes through a linear filter with transfer function $H(\omega)$ the output process Y has spectral density

$$\hat{R}_{YY}(\omega) = |H(\omega)|^2 \hat{R}_{XX}(\omega) \tag{5.13}$$

A random process is said to be *ergodic* if its time average is equal to its statistical average. All the processes we deal with are ergodic.

From Figure 5.7, the output signal at oscillations is

$$y(t) = n(t) + v(t) \tag{5.14}$$

where $y(t)$ is an RV from an ergodic random process Y.

First we note that the transfer function $H(\omega)$ of a feedback network of the type described in (5.8) (in fact of every band-pass network) can be equivalently represented by summing the right ω_0 shift and left ω_0 shift of a corresponding low-pass transfer function H_L (Clarke, 1978)

$$H(\omega) = H_L(\omega - \omega_0) + H_L(\omega + \omega_0)$$

$$H_L(\omega) = \frac{1}{(1 + j2Q_L\omega/\omega_0)} \tag{5.15}$$

Now, we claim that

$$\hat{R}_{VN}(\omega) + \hat{R}_{NV}(\omega) = 0, \omega \neq \omega_0 \tag{5.16}$$

The proof of (5.16) is by no means straightforward, and we defer it to the end of the discussion. For the moment we assume it true. The autocorrelation of the output signal is

$$R_{YY}(\tau) = E[\{v(t) + n(t)\}\{v(t+\tau) + n(t+\tau)\}]$$
$$= R_{VV}(\tau) + R_{NN}(\tau) + R_{VN}(\tau) + R_{NV}(\tau) \tag{5.17}$$

For the thermal noise

$$R_{NN}(\tau) = \frac{N_0}{2}\delta(\tau) \tag{5.18}$$

From Figure 5.7, with $y(t)$ defined in (5.14) and denoting the convolution operator by *, we conclude that

$$v(t) = (y * h)(t) \equiv \int_{-\infty}^{\infty} y(\xi)h(t-\xi)\,d\xi \tag{5.19}$$

The random function $v(t)$ is the result of the random function $y(t)$ going through the linear filter with impulse response $h(t)$. Since $v(t)$ is a sample function from some random process V, it follows from (5.13) that

$$\hat{R}_{VV}(\omega) = |H(\omega)|^2 \hat{R}_{YY}(\omega) \tag{5.20}$$

But in virtue of (5.16), it follows from (5.17) and (5.18) that for $\omega \neq \omega_0$

$$\hat{R}_{VV}(\omega) = \hat{R}_{YY}(\omega) - \frac{N_0}{2} \tag{5.21}$$

Substituting (5.21) into (5.20) yields

$$\hat{R}_{YY}(\omega) = \frac{N_0}{2}\frac{1}{1 - |H(\omega)|^2} \tag{5.22}$$

Noting that $H(\omega)$ is narrowband, we may neglect the influence of $H_L(\omega + \omega_0)$ on the positive frequencies, an we may approximate (5.15) by

$$H(\omega) \approx \frac{1}{1 + j2Q_L\dfrac{\omega - \omega_0}{\omega_0}}, \quad \omega > 0 \tag{5.23}$$

Using (5.23), after a short manipulation of (5.22), we get

$$\hat{R}_{YY}(\omega) \approx \frac{N_0}{2}\left[1 + 1/\left(2Q_L\frac{\omega - \omega_0}{\omega_0}\right)^2\right], \quad \omega > 0, \ |\omega| \neq \omega_0 \tag{5.24}$$

Now, since the energy of $y(t)$ is concentrated near the center frequency ω_0, using the quadrature narrowband noise representation described in Schwartz (1990), $y(t)$ may be approximated as

$$y(t) \approx [V + a(t)] \cos \omega_0 t - b(t) \sin \omega_0 t \qquad (5.25)$$

where V is the amplitude of a noiseless carrier, and $a(t)$ and $b(t)$ are low-pass, zero-mean and uncorrelated *RV*. Now, if the variance of the interfering noise is small compared to the carrier, namely

$$\{Var(a(t)), Var(b(t))\} \ll V^2/2 \qquad (5.26)$$

then, we may further approximate (5.25) to the form

$$y(t) \approx V\{\cos \omega_0 t - [b(t)/V] \sin \omega_0 t\} \qquad (5.27)$$

Equation (5.27) is the well-known approximation to a narrowband FM signal (Clarke, 1978)

$$y(t) \approx V \cos[\omega_0 t + b(t)/V], \quad |b(t)|/V \ll 1 \qquad (5.28)$$

Thus the spectral density (5.24) away from the central frequency ω_0 is due to narrowband FM modulation.

Setting $N_0 = kTF$, dividing the spectral density by the oscillating signal power P_S at limiting port (the input of the active device in Figure 5.7), taking the logarithm, and labeling the result by $L(\omega - \omega_0)$, leads to (5.10).

All we are left with is to prove claim (5.16). First inserting (5.14) into (5.19) we get

$$v(t) = ([n + v] * h)(t) \qquad (5.29)$$

Now, substituting $v(t)$ as defined in (5.29) into the right-hand side of (5.29) yields

$$v(t) = ([n + \{n + v\} * h] * h)(t) = [n * h + n * h * h + v * h * h](t) \qquad (5.30)$$

Substituting again M times in the right-hand side of (5.29) for v, with $M \to \infty$, and denoting by $(*)^m$ the m-fold convolution operator. i.e.

$$n(*)^2 h \equiv n * h * h, \quad n(*)^3 h \equiv n * h * h * h, \ldots \ldots$$

and so on, we get

$$v(t) = \lim_{M \to \infty} \left[n \sum_{m=1}^{M} (*)^m h + v(*)^M h \right](t)$$

$$= \lim_{M \to \infty} \left[n * \left(\sum_{m=0}^{M-1} h(*)^m h \right) + v * \left(h(*)^{M-1} h \right) \right](t) \qquad (5.31)$$

It is easy to see that

$$\lim_{M \to \infty} [v * (h(*)^{M-1}h)](t) = 0 \tag{5.32}$$

Indeed denoting by $F\{x\}$ the Fourier transform of $x(t)$, and recalling the property

$$F\{x * y\} = F\{x\} \cdot F\{y\} \tag{5.33}$$

it follows that

$$F\{h(*)^{M-1}h\} = [H(\omega)]^M \tag{5.34}$$

Since

$$|H(\omega)| < 1, \ \omega \neq \omega_0, \ |H(\omega_0)| = 1 \tag{5.35}$$

it follows that

$$\lim_{M \to \infty} [H(\omega)]^M = \begin{cases} 0, \omega \neq \omega_0 \\ 1, \omega = \omega_0 \end{cases} \tag{5.36}$$

Thus, $[H(\omega)]^\infty$ is a filter of unit amplitude and zero width, so any signal passing through it vanishes.

Alternatively we can state the following: the support of $[H(\omega)]^\infty$ is of zero measure, and thus, when performing the inverse Fourier transform of (5.36), the result vanishes. As a result, (5.31) reduces to the form

$$v(t) = \left[n * \left(\sum_{m=0}^{\infty} h(*)^m h \right) \right]_{(t)} \tag{5.37}$$

Since the operations of expectation and integration are interchangeable (Papoulis, 1991), using (5.37) and (5.18), we get, setting $u = t + \tau$

$$R_{NV}(\tau) = E[n(t)v(t + \tau)] = E[v(u)n(u - \tau)]$$

$$= \int_{-\infty}^{\infty} E[n(\xi)n(u - \tau)] \left(\sum_{m=0}^{\infty} h(*)^m h \right)_{(u-\xi)} d\xi$$

$$= \frac{N_0}{2} \int_{-\infty}^{\infty} \delta(u - \tau - \xi) \left(\sum_{m=0}^{\infty} h(*)^m h \right)_{(u-\xi)} d\xi = \frac{N_0}{2} \left(\sum_{m=0}^{\infty} h(*)^m h \right)_{(\tau)} \tag{5.38}$$

By property (5.34), and by virtue of (5.35), the Fourier transform of (5.38), for $\omega \neq \omega_0$, is the sum of a convergent geometric series.

Using the summation formula $\sum_{n=0}^{\infty} q^n = 1/(1 - q)$, $|q| < 1$ we obtain

$$\hat{R}_{NV}(\omega) = \frac{N_0}{2} \sum_{m=1}^{\infty} (H(\omega))^m = \frac{N_0}{2} \left[\sum_{m=0}^{\infty} (H(\omega))^m - 1 \right]$$

$$= \frac{N_0}{2} \left[\sum_{m=0}^{\infty} \left(\frac{1}{(1 + j2Q_L(\omega - \omega_0)/\omega_0)} \right)^m - 1 \right]$$

$$= \frac{N_0}{2} \frac{1}{j2Q_L(\omega - \omega_0)/\omega_0}, \quad \omega \neq \omega_0 \tag{5.39}$$

Therefore $\hat{R}_{NV}(\omega)$ is purely imaginary

$$Re(\hat{R}_{NV}(\omega)) = 0, \quad \omega \neq \omega_0 \tag{5.40}$$

Noting from (5.38) that

$$R_{NV}(\tau) = R_{VN}(-\tau) \tag{5.41}$$

and since $v(t)$ and $n(t)$ are real-valued

$$\hat{R}_{NV}(\omega) = F\{R_{VN}(-\tau)\} = conjugate(\hat{R}_{VN}(\omega)) \tag{5.42}$$

Thus $\hat{R}_{VN}(\omega)$ is the complex conjugate of $\hat{R}_{NV}(\omega)$. But since $\hat{R}_{NV}(\omega)$ is purely imaginary, it follows that

$$\hat{R}_{VN}(\omega) + \hat{R}_{NV}(\omega) = 0, \quad \omega \neq \omega_0 \tag{5.43}$$

which proves (5.16).

5.4 Bipolar Oscillators

The overwhelming majority of high-performance RF oscillators is based on bipolar transistor design. The reasons are many, and among them: The analytic and mathematical treatment of a bipolar junction is precise and well established, and leads to an actual performance extremely well correlated to the theoretical design.

In oscillator applications, bipolar transistors outperform all other devices as far as performance stability and repeatability properties are concerned. Excellent VCO phase noise, power output and tunability can be achieved with low current consumption, through wide temperature and supply voltage ranges. Integration is easily done, yet allowing much a posteriori design freedom. Oscillator frequencies ranging from tens of kHz and up to 10 GHz can be obtained using nearly the same design approach, thus covering most modern communication applications.

5.4.1 Non-Saturating Bipolar Theory

We summarize here, without proof, the relevant outcome of the theory of non-saturating bipolar transistor behavior under RF signal drive. The interested reader may find the details in Clarke (1978) and Millman-Halkias (1972).

Although we discuss in detail the bipolar case, the general design approach is the same for any two-port nonlinear device with large-signal asymptotic behavior,

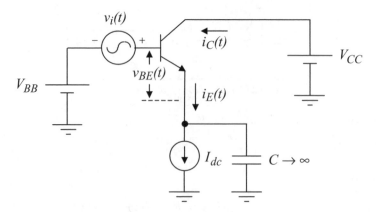

Figure 5.8   Bipolar transistor with bias and RF drive

as it will be shown later in section 5.5.2, when treating the design of CMOS inverter-driven crystal oscillators.

Figure 5.8 shows the synthetic configuration of a bipolar transistor with ideal (zero-impedance) collector and base DC voltage bias sources (V_{BB} and V_{CC}), and a constrained fixed ideal (infinite impedance) DC emitter current source I_{dc}. The base-emitter junction is driven by an RF signal $v_i(t)$, whose positive direction is arbitrarily taken as shown. The total (DC + AC) collector and emitter currents are denoted by $i_C(t)$ and $i_E(t)$ respectively. The resulting total (DC + AC) base-emitter voltage is denoted by $v_{BE}(t)$. The capacitor C is a perfect short circuit at RF frequency and an open circuit at DC.

In all the applications of interest for us, we assume (and double check) that the design is such that the transistor may reach instantaneous cut-off, but that either collector saturation or reverse junction breakdown never occurs. This is a central point in designing bipolar VCOs. If the transistor is allowed to saturate or break down even for a small percentage of the cycle, the oscillator will still work apparently properly, but we will end up with an inexplicable degradation in noise performance or bad repeatability.

The computations to follow are all in MKS units. Denoting by q the electron charge, by k the Boltzmann constant and by $T = 300$ K the room temperature, we define

$$V_T \equiv kT/q \approx 26\ mV \tag{5.44}$$

which is the junction potential at room temperature. By superposition

$$v_{BE}(t) = v_i(t) + V_{dc} \tag{5.45}$$

where V_{dc} is the base-to-emitter DC voltage, which, we anticipate, is strongly affected by the RF drive level.

Then, denoting by I_{ES} the emitter-to-base reverse saturation current, and provided that neither collector saturation nor reverse junction breakdown occur, the total emitter current is given by

$$i_E(t) = I_{ES} e^{v_{BE}(t)/V_T} \tag{5.46}$$

and, for all practical purposes, we assume that the collector and the emitter currents are roughly equal, namely

$$i_C(t) \approx i_E(t) \tag{5.47}$$

Representing for convenience the RF signal at frequency ω by an unmodulated carrier of peak value V

$$v_i(t) = V \cos \omega t \tag{5.48}$$

we denote by x the ratio between the peak signal amplitude and the junction potential

$$x \equiv V/V_T \tag{5.49}$$

Substituting (5.44), (5.45) and (5.48) into (5.46), and expanding the latter in a Fourier series, the total emitter current can be represented in the form

$$i_E(t) = I_{dc} \left[1 + 2 \sum_{n=1}^{\infty} \frac{I_n(x)}{I_0(x)} \cos(n\omega t) \right] \tag{5.50}$$

where $I_n(x)$ is the modified Bessel function of order n and argument x

$$I_n(x) \equiv \frac{1}{2\pi} \int_{-\pi}^{\pi} e^{x \cos \theta} \cos(n\theta) \, d\theta \tag{5.51}$$

and the constrained constant DC emitter current source I_{dc} satisfies

$$I_{dc} = I_{ES} e^{V_{dc}/V_T} I_0(x) \tag{5.52}$$

The modified Bessel functions have asymptotic behavior for large or small values of x. In particular we have the asymptotic expansions

$$I_n(x) \approx \frac{(x/2)^n}{n!}, \quad x \ll 1 \tag{5.53}$$

$$I_n(x) \approx \frac{e^{x - \frac{n^2}{2x}}}{\sqrt{2\pi x}}, \quad x \gg 1 \tag{5.54}$$

Thus implying

$$I_0(x) \approx 1, \quad x \ll 1 \tag{5.55}$$

$$\frac{I_n(x)}{I_0(x)} \approx e^{-\frac{n^2}{2x}}, \quad x \gg 1 \tag{5.56}$$

In fact, as we see later, robust oscillator design will always end up with large values of x. It is worth noting that (5.50) and (5.56) imply that, when $x \to \infty$, all the harmonics in the collector current (including fundamental) approach the same peak amplitude value of $2I_{dc}$. This is an important feature allowing the design of harmonic oscillators and frequency multipliers, as well as being a major contributor to design stability.

Since I_{dc} and V_T are constant, then (5.52) shows that the base-emitter DC voltage V_{dc} must change when the amplitude of the RF signal changes. Thus we split the DC voltage across the base-emitter junction into two components as follows:

$$V_{dc} = V_{dcQ} - \Delta V \tag{5.57}$$

where V_{dcQ} is the quiescent (with no RF signal applied) base-emitter DC voltage, and the 'depression voltage' ΔV is a positive value denoting the base-emitter DC voltage change due to the presence of the RF drive.

A depression voltage phenomenon will show up in any device that exhibits clamping of the RF signal at its input (such as JFET). As usual, for silicon transistors we assume $V_{dcQ} \approx 0.7$ V.

Equation (5.57) suggests that, with RF applied, we should expect the base-emitter DC voltage to diminish. In fact, for large drive, we anticipate that the base-emitter junction may even appear reverse biased, while in fact the transistor will be active.

Setting $x = 0$ in (5.52), and using (5.55), we obtain

$$V_{dcQ} = V_T \ln \frac{I_{dc}}{I_{ES}} \tag{5.58}$$

For large values of x, the depression voltage ΔV is a precise measure of the peak RF amplitude across the base-emitter junction.

In particular, substituting (5.57) and (5.58) into (5.52), and using (5.54) for large x and $n = 0$, we obtain

$$e^{\frac{\Delta V}{V_T}} = \frac{e^x}{\sqrt{2\pi x}} \tag{5.59}$$

Taking the natural logarithm of (5.59), noting that $\ln\sqrt{2\pi x} \ll x$ for large x and using (5.49) we finally obtain

$$\Delta V \approx V \tag{5.60}$$

Equation (5.60) provides important means for measuring the peak RF level of a high-frequency oscillator, just by measuring the apparent DC bias.

Indeed the peak base-emitter RF amplitude is nearly equal to the difference between the quiescent value $V_{dcQ} \approx 0.7$ V and the measured base-emitter DC voltage during oscillations. If the oscillation voltage peak is greater than 0.7 V, then the junction will appear reverse-biased.

Substituting (5.45), (5.48) and (5.52) into (5.46) yields the instantaneous emitter current in the form

$$i_E(t) = \frac{I_{dc}}{I_0(x)} e^{x \cos \omega t} \tag{5.61}$$

Equation (5.61) explains why, for large x, all the harmonic components of the emitter current approach the same value.

Indeed, as $x \to \infty$, the current $i_E(t)$ has strong peaks when $\omega t = 2\pi m$, $m = 0$, $\pm 1, \pm 2, \ldots$, and approaches a train of Dirac time impulses, whose Fourier transform is a train of Dirac frequency impulses. A similar behavior will occur in any nonlinear device for which the conduction angle decreases while increasing the input RF drive.

Substituting (5.54) for large x and $n = 0$, (5.61) yields the peak emitter current

$$i_E|_{peak} = I_{dc}\sqrt{2\pi x}, \quad x \gg 1 \tag{5.62}$$

Equation (5.62) warns us that the instantaneous peak emitter current is much larger that the DC current, thus care should be taken when designing the collector load.

For all harmonic components, the collector load should present an impedance low enough to prevent instantaneous collector saturation, otherwise (5.46) will not hold and the all analysis done up to here will be incorrect.

Substituting (5.56) into (5.50) we see that the nth emitter harmonic current $i_{E,n}(t)$ is given by

$$i_{E,n}(t) = 2I_{dc} \cos(n\omega t) \cdot [1 + O(n^2/2x)] \tag{5.63}$$

from which we reach the important conclusion that, for $x \to \infty$, the peak amplitude of the harmonic components of the emitter current is *independent* of the value of x, and is equal to twice the transistor DC bias current.

$$\lim_{x \to \infty} i_{E,n}(t) = 2I_{dc} \cos(n\omega t), \quad n = 1, 2, 3, \ldots \tag{5.64}$$

Indeed, in practice, the fundamental and the first few harmonic components of the emitter current stabilize near a peak value of $2I_{dc}$ for moderate values of x and change very little with variations of x. Thus, as pointed out before, designing for large x will yield a robust, reliable and repeatable oscillator design.

5.4.2 Detailed Bipolar VCO Design

In order to ease the understanding of the (pretty complex) VCO design process, we explain the theory in the framework of a complete computational design example including all the details, from requirements to full VCO scheme. For background and insight on physical bipolar transistor parameters refer to Millman-Halkias (1972).

We design a VCO to be incorporated into the synthesizer of a VHF (around 150 MHz) radio with channel spacing of 25 kHz (a real-life case).

The requirements for the VCO are:

- Center frequency: $f_0 = 150$ MHz.
- *SSB* noise at adjacent channel: $L(f_0 \pm 25$ kHz$) < -118$ dBc/Hz.
- Tunability: $f_{max} - f_{min} \geq 2$ MHz over a control voltage swing <5 V.
- Output power delivered to load: ≥ 0 dBm on 50 Ω.
- DC supply voltage: 10 V.
- DC current consumption: ≤ 5 mA.

The figures of the bipolar transistor are:

- Base-spreading resistance: $r_{bb'} = 20$ Ω.
- Noise figure: $NF = 2$ dB.
- DC current gain $\beta = 80$.
- Base-emitter reverse breakdown voltage: 1.8 V.
- Collector-base transition capacitance: $C_{cb} = 1$ pF @ $V_{cb} = 5$ V.
- Gain-bandwidth product: $f_T = 6$ GHz.

The design will sequentially follow the steps below:

1. Determine the components (including Z_1 and Z_3 in Figure 5.2).
2. Determine the required resonator figures (Z_2 in Figure 5.2).
3. Design the resonator to achieve power out, noise and tunability.

The configuration we chose is the Colpitts architecture, which is very convenient for stable BJT implementation.

Figure 5.9 shows the VCO circuit, where the resonator characteristics are still to be determined based on the required performance. The VCO RF load is not shown, however, the analysis takes into account its effects. The reason why we postpone the RF output issue is that coupling the VCO to its load is not straightforward, and we will deal with it in detail later on.

The dotted box in Figure 5.9 represents the resonator equivalent. In particular:

- X_R is the serial (inductive) reactance of the resonator at oscillating frequency.
- r is the serial resistance of the resonator, representing the resistive losses due to the resonator itself and the effect of the power delivered to load, but not including the losses due to the transistor and associated circuitry.
- The RF choke RFC is an open circuit at RF frequencies, and a short circuit at DC.

As explained in section 5.1, as far as the RF behavior is concerned, the Colpitts topology and the π topology of Figure 5.3 are equivalent. Thus, instead of analyzing Figure 5.9 we may equivalently look at Figure 5.10. Now, let us represent the actual transistor as an ideal transistor, with lumped stray components as shown in Figure 5.11.

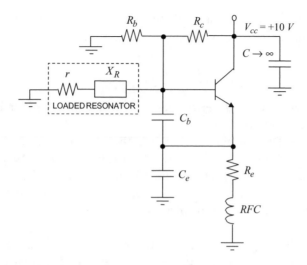

Figure 5.9 Colpitts bipolar VCO

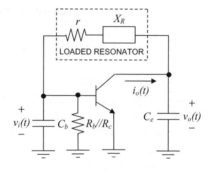

Figure 5.10 π-topology equivalent of the Colpitts bipolar VCO

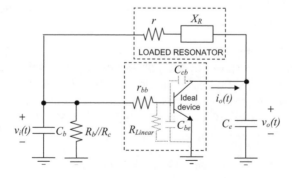

Figure 5.11 Lumped representation of the transistor in Figure 5.10

With I_{dc} denoting the transistor DC bias current, we compute the base-emitter diffusion capacitance C_{be} as (Millman-Halkias, 1972):

$$C_{be} \approx \frac{I_{dc}}{2\pi f_T V_T} \tag{5.65}$$

Assuming $x \gg 1$, we will verify that the fundamental emitter RF current has a peak amplitude value of about $2I_{dc}$.

Then, since the DC current flowing into the ideal base is I_{dc}/β and it is the average value resulting from the flow of narrow current pulses just as in the emitter, we deduce that the fundamental RF current flowing into the base of the ideal transistor has peak amplitude $I_1 = 2I_{dc}/\beta$ (as anticipated in section 5.4.1, the first few harmonics of a train of narrow impulses have about twice the DC value).

If we assume that Q_L, the loaded Q of the π network, is large, then $v_i(t)$ is a pure sinusoidal voltage of peak amplitude V at the fundamental frequency.

In this case, it can be shown (Clarke, 1978) that the equivalent linear resistive base-emitter junction load R_{Linear} seen by $v_i(t)$ is approximately equal to V/I_1 (the ratio of the peak fundamental voltage to the peak fundamental current), namely

$$R_{Linear} = \frac{\beta V_T}{2I_{dc}} x \tag{5.66}$$

Equation (5.66) is a basic result stating that as the drive voltage increases, the value of the equivalent linear loading resistor gets larger. Since I_{dc} and x have not yet been determined, for the moment we cannot compute either R_{Linear} or C_{be}.

With reference to (5.47), (5.48), (5.49) and (5.64) and as just noted above, if the oscillator is designed to stabilize to an RF oscillation level such that $x \gg 1$, then the fundamental collector/emitter current is virtually independent from x. From (5.50) and (5.56), we see that the fundamental emitter current is

$$i_{E,1}(t) = 2I_{dc}\frac{I_1(x)}{I_0(x)}\cos(\omega_0 t) \approx 2I_{dc}e^{-\frac{1}{2x}}\cos(\omega_0 t) = 2I_{dc}\cos(\omega_0 t) + O\left(\frac{1}{2x}\right) \tag{5.67}$$

We anticipate that, in most cases, both the attenuation and the phase shift introduced by the combination of the base-spreading resistance and the diffusion capacitance in Figure 5.11 can be neglected. It follows that, even for $x \geq 5$, roughly corresponding to a peak base-emitter RF signal of $V \geq 130$ mV, and no matter what the value of x, with the help of (5.67), and with reference to Figure 5.8, we may approximate in Figure 5.11

$$v_i(t) = x V_T \cos(\omega_0 t) \Rightarrow i_o(t) \approx -2I_{dc}\cos(\omega_0 t), x \geq 5 \tag{5.68}$$

At this point we introduce, without proof, the following well-known two-way serial-parallel RC conversion formula (5.69) (Krauss, Bostain and Raab, 1980) described in Figure 5.12 (the proof is left as an exercise to the reader).

Figure 5.12   RC serial-parallel conversion

The formula states that in order to convert a serial RC configuration to an equivalent parallel one (or the converse), we should first compute Q for the given configuration, and then use the formula including r and R to compute the resistor conversion

$$Q_S = \frac{1}{\omega_0 r C_S}, \quad Q_P = \omega_0 R C_P, \quad R = (Q_S^2 + 1)r, \quad r = R/(Q_P^2 + 1)$$

$$Q \gg 1 \Rightarrow C_S \approx C_P, \quad \{Q_S, Q_P\} = Q, \quad R \approx Q^2 r \tag{5.69}$$

If $Q \gg 1$, the capacitor remains unchanged, while the resistor is much smaller in the serial topology than in the equivalent parallel one.

It follows from (5.65), (5.66) and (5.69) that, once I_{dc} and x are determined, we will be able to reflect all the cumulative loading effects of r, $r_{bb'}$, R_b, R_c and R_{Linear} in the form of a single serial resistor r_{load} which will replace r in Figure 5.11, and will determine the value of Q_L, the overall loaded Q of the resonant circuit.

Thus our next task is to determine the required values of I_{dc} and x. Assume that the effects of C_{cb} (the collector-base transition capacitance) and C_{be}, have been incorporated into C_b and C_e (we will see how to do that shortly), and denote the resulting capacitors by C_b' and C_e'. Then the equivalent circuit we work with, from now on, is the one shown in Figure 5.13.

From (5.5) and (5.64), and for $x \gg 1$, the transfer function of the π network at ω_0 yields

$$v_i(t) = -\frac{1}{\omega_0^2 C_e' C_b' r_{load}} i_o(t) = \frac{2 I_{dc}}{\omega_0^2 C_e' C_b' r_{load}} \cos(\omega_0 t) \tag{5.70}$$

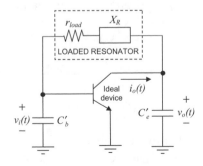

Figure 5.13   All-inclusive equivalent oscillator circuit

from which it follows that

$$x = \frac{2I_{dc}/V_T}{\omega_0^2 C_e' C_b' r_{load}} \tag{5.71}$$

Equation (5.71) is a fundamental result, stating that for large values of x, the amplitude at which the bipolar oscillator stabilizes is directly proportional to the DC bias current, and inversely proportional to the loaded resonator losses, but virtually independent from anything else, including transistor variations. This is why the design around large values of x leads to stable and repeatable performance.

In light of the phase noise analysis previously done, we would like to design the oscillator so as to obtain the largest possible value of x, as this would imply the best noise floor performance. Indeed, for a given bias current, and given resonator losses, the value of x can be maximized to a value x_{max}.

Using (5.4) and setting $C_b' = kC_e'$ we get

$$\omega_0 = \frac{\left(1 + \dfrac{1}{k}\right)}{L(\omega_0)C_e'}, \quad k = \frac{C_b'}{C_e'} \tag{5.72}$$

Substituting (5.72) into (5.71), differentiating x with respect to k and equating to zero, we find that the value of x is maximal when we take $k = 1$, namely $C_b' = C_e'$

$$C_b' = C_e' \Rightarrow x = x_{max} \tag{5.73}$$

Using (5.69), we may equivalently represent the losses in the form of a single parallel or serial resistor located at any arbitrary point of the π network.

In particular, when $C_b' = C_e'$, due to symmetry, we may reflect the losses in the form of an equivalent large resistor R_{load} in parallel to either C_b' or C_e', instead of a small resistor r_{load} in series to the resonator. Since the loss must be the same, no matter how we represent it, then the voltage on both C_b' and C_e' must have the same amplitude (and opposite sign due to the negative transfer function). Thus, when x is maximized, the voltages $v_i(t)$ and $v_o(t)$ in Figure 5.13, have both the same amplitude value $x_{max} V_T$, and opposite sign, namely

$$C_b' = C_e' \Rightarrow v_i(t) = x_{max} V_T \cos \omega_0 t = -v_o(t) \tag{5.74}$$

It follows from (5.74), that the base-to-collector large-signal voltage gain G_V at resonance is

$$G_V = -1 \tag{5.75}$$

Using Miller's theorem (Millman-Halkias, 1972), the immediate outcome of (5.75) is that the transition capacitance C_{cb} is equivalent to two identical capacitors of value $2C_{cb}$ each, one in parallel to C_b and the other in parallel to C_e. For the

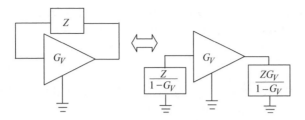

Figure 5.14   Miller theorem. G_V = voltage gain, Z = impedance

sake of completeness we show the Miller equivalence in Figure 5.14, where Z denotes an arbitrary impedance.

From (5.68) and (5.74) the total RF power P_o delivered to the lossy π network of Figure 5.13 is

$$P_o = (v_o|_{rms})(i_o|_{rms}) = \frac{x_{max} V_T}{\sqrt{2}} \frac{2 I_{dc}}{\sqrt{2}} = x_{max} V_T I_{dc} \qquad (5.76)$$

Since we are asked to deliver to the output load a power $P_L \geq 1$ mW, let us aim for

$$P_o = 3 \; mW \qquad (5.77)$$

The reason for our choice in (5.77) provides an important insight: if all the power P_o had been delivered to the resonator itself, we would have had $Q_L = Q_U$, which according to Leeson's equation would achieve the best possible noise performance. If we deliver 1 mW (one-third of the total power) to the load, then Q_L cannot be better than two-thirds of Q_U, since r_{load} becomes larger by one-third. Thus, in order to preserve the order of magnitude, namely to obtain $Q_L = O(Q_U)$, the total delivered power P_o must always be designed substantially larger than the power delivered to load. Delivering power to the loaded resonator is unavoidable, but a good design may lead to negligible losses due to transistor circuitry.

Equation (5.76) shows that in order to reduce the current consumption we must set x_{max} as large as possible. However, it is not possible to increase x_{max} indefinitely. The reason becomes clear by looking at (5.57) and (5.60), which together imply that the average DC voltage across the base-emitter junction is given by the quiescent (non-oscillating) DC voltage, less the oscillation peak. This effect tends to reverse-bias the junction, indeed

$$V_{dc} = V_{dcQ} - V_T x_{max} = 700 - 26 x_{max} \quad [mV] \qquad (5.78)$$

On top of V_{dc} rides the RF signal $v_i(t) = x_{max} V_T \cos(\omega_0 t)$. Thus the maximal instantaneous reverse voltage on the base-emitter junction is

$$V_{be}|_{reverse} = V_{dc} - x_{max} V_T = V_{dcQ} - 2 x_{max} V_T = 700 - 52 x_{max} \quad [mV] \quad (5.79)$$

Since the reverse base-emitter breakdown voltage is 1.8 V, then, in order to avoid instantaneous junction breakdown, we must satisfy

$$V_{be}|_{reverse} > -1.8V \Rightarrow x_{max} < 48 \tag{5.80}$$

More generally, if we denote by V_{break} the absolute value of the reverse base-emitter breakdown voltage, then we must have

$$x < \frac{V_{break} + V_{dcQ}}{2V_T} \tag{5.81}$$

If condition (5.81) is not met, instantaneous breakdown will occur, which will cause nonlinear loading of the resonant network, reducing the loaded Q and leading to an unexplainable degradation in phase noise performance.

A side remark: looking at (5.81) it becomes clear why FET oscillators are able to provide noise performance better than BJT oscillators, and are sometimes used in very demanding designs. Indeed, their reverse junction breakdown voltage is higher, which allows for larger amplitude of oscillation. BJT devices, however, are superior in most other aspects.

For our design we take the fair peak RF oscillation level of 0.78 V, which yields

$$x_{max} = 30 \tag{5.82}$$

Now that P_o and x_{max} are defined, (5.76) calls for an (oscillating) current drain of

$$I_{dc} = \frac{P_o}{x_{max} V_T} \approx 4 \ mA \tag{5.83}$$

To bias the transistor, we arbitrarily chose

$$R_b = R_c = 33 \ k\Omega \tag{5.84}$$

Since by (5.60) and (5.82), due to the oscillations, the base-emitter DC voltage is about zero, and taking into account the finite transistor β, the required emitter resistor per (5.83) turns out to be

$$R_e \approx 1 \ k\Omega \tag{5.85}$$

We are now in position to compute the value of r_{load}.

The first step will be to convert the loading of the transistor and all its components into a resistor in series to the resonator. Then, since the transistor must contribute one-third of the power losses, r_{load} should have a value three times larger. From (5.65) and (5.83)

$$C_{be} \approx \frac{I_{dc}}{2\pi f_T V_T} = 4 \ pF \tag{5.86}$$

and from (5.66), (5.82) and (5.83)

$$R_{Linear} = \frac{\beta V_T}{2I_{dc}} x_{max} = 7.8 \; k\Omega \tag{5.87}$$

Since $1/(\omega_0 r_{bb'} C_{be}) \approx 13 \gg 1$, then $r_{bb'} \ll 1/(\omega_0 C_{be})$, thus $r_{bb'}$ neither reduces appreciably the input voltage to the base emitter junction, nor introduces significant phase shift. Therefore, by (5.75) and using the Miller equivalence on C_{cb}, Figure 5.11 takes the form of Figure 5.15.

Sequentially using the transformation (5.69) on Figure 5.15 we reflect R_{Linear} and $r_{bb'}$ in parallel to C_b, thus R_{Linear}, $r_{bb'}$, R_c and R_e, are transformed into a resistor R' in parallel to C_b. Finally, using (5.69), we convert R' into a serial equivalent resistor r' as shown in Figure 5.16.

Incorporating lumped and stray capacitances into C'_b and C'_e we get (the detailed computation is left to the reader).

$$R' = 1200 \; \Omega, \;\; C'_b = C_b + 6 \; pF, \;\; C'_e = C_e + 2 \; pF \tag{5.88}$$

Since (5.73) requires $C'_b = C'_e$, then (5.88) implies

$$C_e = C_b + 4 \; pF \tag{5.89}$$

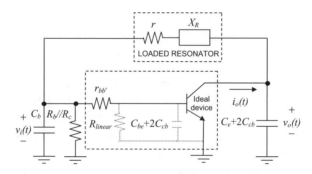

Figure 5.15 Using the Miller effect on Figure 5.11

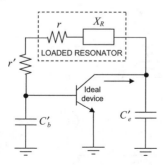

Figure 5.16 Converting R_{Linear}, $r_{bb'}$, R_c and R_e into r'

The value of C_b is chosen based on the following considerations: in order to have a stable value of C_b', the lumped capacitor C_b must have a value at least comparable to the stray capacitances. Thus, since the stray capacitances amount to 6 pF, we pick up an (arbitrary) value

$$C_b = 12 \ pF \Rightarrow C_b' = 18 \ pF \qquad (5.90)$$

from which (5.69) yields

$$r' \approx 3 \ \Omega \qquad (5.91)$$

and from (5.89) it follows

$$C_e = 16 \ pF \qquad (5.92)$$

Substituting (5.73) into (5.71), and noting that

$$r_{load} = r' + r \qquad (5.93)$$

we obtain

$$x_{max} \approx \frac{2I_{dc}/V_T}{\omega_0^2 (C_b')^2 (r'+r)} \qquad (5.94)$$

which, solving for r yields

$$r \approx 35 \ \Omega \qquad (5.95)$$

From which

$$r_{load} = r' + r \approx 38 \ \Omega \qquad (5.96)$$

With $x_{max} = 30$ the transistor output current is independent from the base-emitter voltage, then the transistor input in Figure 5.16 *is effectively disconnected* from its output, and we may view the complete circuit in the form of Figure 5.17, where the current source is independent and, at resonance, is directly coupled to the resistor R_{load}, which is the parallel transformation of r_{load} using (5.69).

Denote by r_R and r_L the contribution, within r_{load}, of resonator and load dissipation respectively, namely

$$r_R + r_L = r_{load} - r' \qquad (5.97)$$

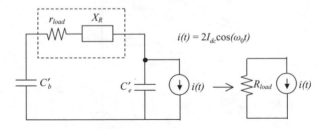

Figure 5.17   Equivalent of Figure 5.16 at resonance

With a peak current of $2I_{dc}$ the total power dissipated in r_{load} is

$$P_o = I_{dc}^2 R_{load}, \quad R_{load} \approx (\omega_0 r_{load} C_e')^{-2} r_{load} \tag{5.98}$$

The (load) resistor r_L must dissipate a power $P_o/3$, then (5.97) implies that

$$r_L = r_{load}/3 \Rightarrow r_R = r_{load}(1 - 1/3 - r'/r_{load}) \approx 0.59 \, r_{load} \tag{5.99}$$

and thus by (5.98), the resonator dissipates $0.59 P_o$. It follows that, no matter what kind of resonator we choose, we end up with

$$Q_L = \frac{r_R}{r_{load}} Q_U \approx 0.59 \, Q_U \tag{5.100}$$

The minimal value of Q_U to achieve the required phase noise performance is readily inferred from the Leeson's equation (5.10).

It is straightforward to verify that the power dissipating into the transistor input is

$$P_S = P_o r'/r_{load} \cdot R'/R_{linear} \approx 0.04 \, mW \tag{5.101}$$

where R' is the equivalent of r' in parallel to C_b', computed with (5.69).

With the requirement for $L = -118$ dBc/Hz at 25 kHz from the center frequency of 150 MHz, (5.10) yields

$$10^{-11.8} = \frac{1}{2} \cdot \frac{1.38 \cdot 10^{-23} \cdot 300 \cdot 10^{0.2}}{0.04 \cdot 10^{-3}} \times \left[1 + 1 \bigg/ \left(2Q_L \frac{25 \cdot 10^3}{150 \cdot 10^6} \right)^2 \right]$$

$$\Rightarrow Q_L \approx 22 \tag{5.102}$$

from which it follows that we must satisfy

$$Q_U \geq Q_L/0.59 \approx 37 \tag{5.103}$$

At this point, using (5.96), (5.97), (5.99), (5.103) and Figure 5.17, we are in position to completely specify the resonator performance.

At $\omega = \omega_0$ the loaded/unloaded resonator equivalent must exhibit the following characteristics:

$$Q_U \geq 37, \quad Q_L = 22$$

$$r_R \approx 22 \, \Omega, \quad r_L \approx 13 \, \Omega$$

$$X_R = 1/j\omega_0 C_T$$

$$C_T = C_e' C_b'/(C_e' + C_b') \tag{5.104}$$

and, for convenience we summarize below the circuit components:

$$C_b = 12 \, pF, \quad C_e = 16 \, pF$$

$$R_b = R_e = 33 \, k\Omega$$

$$V_{cc} = 10 \, V \tag{5.105}$$

Equation (5.104) bears the reason why X_R cannot be implemented by simply using an inductor (whether lumped or distributed). Indeed the following independent requirements must be met:

1. In order to satisfy (5.104), we must be able to control the values of r_R and r_L without affecting the values of Q_U and Q_L. In other words, the resonator must be capable to perform an impedance transformation and exhibit resistive values at our request.
2. The resonator must exhibit a specific inductive reactance value X_R, which is dictated by circuit components.
3. The topology of the resonator must allow for the required tuning range, usually by means of a varactor, while keeping the required unloaded Q (the varactor is considered as being a part of the resonator, and has the effect of lowering the value of Q_U).

NOTE: The power to load is usually extracted from the resonator itself for reasons of spectral purity and impedance control.

It is possible to drive the output power from other points in the circuits, but then the output signal will be rich in harmonic content, and additional buffering and filtering will be needed.

As can be easily appreciated by looking at (5.64), since for large oscillation amplitudes all the harmonic currents tend to the same value, we may extract harmonic components of the collector current instead of the fundamental, thus obtaining an harmonic oscillator which provides frequency multiplication 'for free'.

The architecture of choice for implementing the impedance-transforming resonator functionality in most lumped realizations, is the parallel, lightly-coupled resonant circuit in Figure 5.18, where R_p represents the losses due to inductor, load, and tuning varactor.

The circuit can be shown to be equivalent to a serial tuned circuit whose impedance and equivalent values are described in (5.106). Zero-pole diagram techniques for obtaining the approximated expressions below, are detailed in Clarke (1978). The resonator can be equivalently realized by using a lightly-coupled tuned transmission line of electrical length $\approx \lambda/4$.

Here we bring the detailed computation for the lumped circuit, then, we can obtain lumped equivalents from every type of transmission line, and build in this

Figure 5.18   Impedance-transforming resonator architecture and its equivalent

way distributed resonators at high frequencies using the very same procedure. We show later on how to carry out this task. The interested reader may find more insight in Jordan (1989). The equivalence relations for Figure 5.18 are

$$Z_{in} \approx r_s \left[1 + j2Q_s \frac{\omega - \omega_0}{\omega_0} \right], \ Q_s \gg 1, C_c/C_p \ll 1$$

$$\omega_0 = 1/\sqrt{(1 + C_c/C_p)L_p C_p}, \ Q_s = \omega_0 R_p C_p (1 + C_c/C_p)$$

$$r_s = \frac{C_p + C_c}{C_c^2 \, \omega_0 Q_s}, \quad C_s = \frac{C_c^2}{C_p + C_c}, \quad L_s = \left(\frac{C_p + C_c}{C_c} \right)^2 L_p \qquad (5.106)$$

It is easy to see that if the circuit of Figure 5.18 is used as the resonator in the Colpitts configuration of Figure 5.9, we end up with the π topology of Figure 5.13. The advantage of this choice now becomes apparent: from (5.106) it follows that we are able to obtain a wide range of values for r_s, while affecting very little the remaining parameters, thus we can simultaneously satisfy (5.8) and (5.95). We would not been able to achieve this by using just an inductor for X_R.

Now, with all the energy-dissipating elements taken into account in r_{load}, as far as losses are concerned, we may remove the transistor and associated resistors from Figure 5.9, and the resonant circuit looks as in Figure 5.19. where C_e' is still to be determined.

From Figure 5.19 and recalling that, for the fully loaded resonator, (5.102) requires at least $Q_L = 22$, and for maximal oscillation voltage $C_b' = C_e'$, it follows that we must satisfy

$$\frac{1}{C_c} = \frac{1}{C_c'} + \frac{1}{C_b'} + \frac{1}{C_e'} = \frac{1}{C_c'} + \frac{2}{C_b'}, \quad Q_s = Q_L = 22 \qquad (5.107)$$

R_p in Figure 5.19 represents *all* the losses, and corresponds to the serial resistor r_{load}.

As far as the inductor (or transmission line) is concerned, the pool of available components satisfying (5.103) at the given frequency is usually limited, and we choose it arbitrarily. In our example we take

$$L_p = 130 \ nH, \quad Q_U = 80 \qquad (5.108)$$

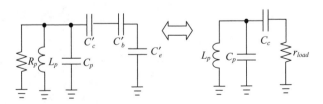

Figure 5.19 Fully loaded, impedance transforming resonant circuit

Now, combining the equations in the second line of (5.106) yields

$$\omega_0 Q_s = R_p / L_p \Rightarrow R_p \approx 2.7 \ k\Omega \tag{5.109}$$

Then, from the equation for Q_s we get

$$C_p + C_c = Q_s / \omega_0 R_p \approx 8.6 \ pF \tag{5.110}$$

Now we may compute C_c using the equation for r_s in (5.106)

$$r_s = r_{load} = 38 = \frac{C_p + C_c}{C_c^2 \ \omega_0 Q_s} \Rightarrow C_c \approx 3.3 \ pF \tag{5.111}$$

From which it follows that

$$C_p \approx 5.3 \ pF \tag{5.112}$$

From (5.107) and (5.90), we get the value of the physical capacitor C_c'

$$\frac{1}{C_c'} = \frac{1}{C_c} - \frac{2}{C_b'} \Rightarrow C_c' \approx 12 \ pF \tag{5.113}$$

Since R_p represents all the power dissipated (3 mW) given in (5.77), it can be seen as a combination of parallel resistors, each one dissipating its own power. In this setting

$$\frac{1}{R_p} = \frac{1}{R_R} + \frac{1}{R_L} + \frac{1}{R_C} \tag{5.114}$$

where R_R, R_L and R_C correspond to the losses due to resonator, load and transistor circuit, respectively.

As long as we take care that the combination of all resistors yields a value equal or higher than the one given in (5.109), the loaded Q will be at least as large as planned. Thus, let us see first how much is dissipated by the inductor. Introducing for the inductor, a transformation similar to (5.69) (the proof is left as an exercise) and described in Figure 5.20 and in (5.115)

$$\frac{\omega_0 L}{r} = Q \gg 1 \Rightarrow Q = \frac{\omega_0 L}{r} \approx \frac{R}{\omega_0 L}, \quad R \approx Q^2 r \tag{5.115}$$

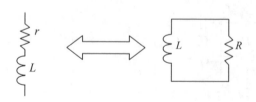

Figure 5.20 LC serial-parallel conversion

We get

$$Q_U = \frac{R_R}{\omega_0 L_p} \Rightarrow R_R \approx 9.8 \ k\Omega \tag{5.116}$$

The transistor circuitry dissipates a power $r'/r_{load} P_o$, thus

$$R_C = (r_{load}/r') R_p \approx 34 \ k\Omega \tag{5.117}$$

Therefore we are able to allocate to the load, a corresponding value

$$\frac{1}{R_L} = \frac{1}{R_p} - \frac{1}{R_R} - \frac{1}{R_C} \Rightarrow R_L \approx 4.1 \ k\Omega \tag{5.118}$$

From which it follows that we can deliver to the load a power P_{load} equal to

$$P_{load} = (R_p/R_L) P_0 \approx 2 \ mW = +3 \ dBm \tag{5.119}$$

A good place to extract the load power is the resonator itself. The reason is that, since $Q_L = 22 \gg 1$, the harmonic content on the inductor–capacitor combination is small, i.e. we have a 'clean' RF signal, which spares further filtering.

The load coupling is achieved in a simple way, by means of a small capacitor C_o, which must be subtracted from C_p as shown in Figure 5.21.

The computation of C_o is straightforward using (5.69), to give

$$R_L = 50/(50 \ \omega_0 C_o)^2 \Rightarrow C_o \approx 2.3 \ pF \Rightarrow C_p' = 3.3 \ pF \tag{5.120}$$

From which

$$C_p' = C_p - C_o = 3.3 \ pF \tag{5.121}$$

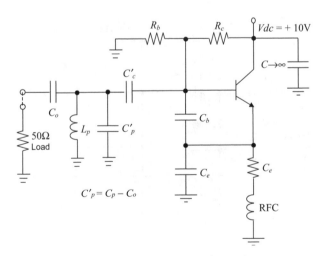

Figure 5.21 Coupling the oscillator

All we are left with is to design the tuning arrangement, which is a straightforward task.

According to (5.110) the total capacitance in parallel to the inductor is $C_t = 8.6$ pF, and since

$$\omega = 1/\sqrt{L_p(C_t + \Delta C)} \approx \omega_0(1 - \Delta C/2C_t) \qquad (5.122)$$

the required 2 MHz tunability implies

$$\Delta C \geq 2C_t \Delta\omega_0/\omega_0 \Rightarrow \Delta C \geq 0.23 \ pF \qquad (5.123)$$

Choosing a common varactor (we neglect varactor dissipation for the sake of simplicity) with capacitance characteristics

$$C_{VAR} = \begin{cases} 40 \ pF \ @0.5V \\ 32 \ pF \ @2.5 \ V \\ 22 \ pF \ @5 \ V \end{cases} \qquad (5.124)$$

and since according to (5.121) $C_p' = 3.3$ pF, we substitute in lieu of C_p' a capacitor C_q in series with the varactor.

To obtain the required capacitance variations, using (5.124), we require

$$\frac{1}{C_q} + \frac{1}{40 \ pF} < \frac{1}{3.415 \ pF}$$

$$\frac{1}{C_q} + \frac{1}{22 \ pF} > \frac{1}{3.185 \ pF}$$

which implies

$$3.72 \ pF < C_q < 3.73 \ pF$$

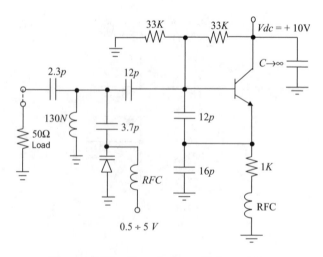

Figure 5.22   Complete oscillator scheme

and we pick up the standard value

$$C_q = 3.7 \ pF \tag{5.125}$$

The complete oscillator scheme is shown in Figure 5.22.

5.5 Crystal Oscillators

In modern equipment, crystal oscillators are used almost exclusively for generating clock signals for synthesizers, digital circuits and processors. The active portion is usually part of a processor or some logic IC, therefore, we will treat only oscillators whose active circuitry consists of a CMOS inverter gate. The theory presented, however, is directly extendable to any type of active device. Since crystals are very well treated in the literature, we show known results without proof.

A word of warning: crystal oscillators driven by logic gates look easy to implement. Even if the circuit is badly designed, or completely miscalculated, in most cases it will oscillate and look fine at first sight. For this reason, engineers often tend to 'guess' the design rather than compute it. However, the punishment will show up in the form of erratic start up, distorted output waveforms, oscillation stop under voltage, temperature or component variations, crystal damage over time, etc.

5.5.1 Piezoelectric Crystal Outlook

We do not treat in detail the theory of piezoelectric crystals, as it is not needed for the actual oscillator design. We deal only with fundamental-mode design (making the crystal oscillate at the lowest possible resonating frequency), which is the most practical one, and we refer the reader to the literature for overtone mode (making the crystal oscillate at odd harmonics of the fundamental frequency), which is used very seldom in modern design. In fact, overtone mode often does not pay off, as it greatly complicates oscillator design and is prone to parasitic effects and spurious oscillations that require much care to be controlled.

For our purposes, a fundamental-mode piezoelectric crystal, driven by a CMOS inverter, can be brought to a form equivalent to a serial tuned circuit operating in a topology identical to the one employed in Figure 5.5.

A few important remarks follow:

THE GOOD NEWS

- The unloaded quality factor Q_U of a crystal resonator is extremely high, with values of the order of $>10,000$, while lumped resonators can hardly achieve a Q_U of 100.
- The high value of Q_U leads to oscillator designs that are extremely stable with respect to variations of the values of external components.

The resonant frequency variations of a crystal over temperatures ranging from $-30\,°C$ to $+60\,°C$ and over aging, are of the order of few PPM (parts per million), while lumped resonators experience variations of several thousands of PPM.

THE BAD NEWS

- The good stability has a drawback: the resonating frequency of a crystal can be tuned only a little by means of external components such as varactor diodes. In fact the frequency can be 'pulled' only a few tens PPM away from its natural value. Crystal oscillators fit only single-frequency applications such as clocks.
- Piezoelectric crystals may dissipate powers of the order of 0.1 mW at most. Above that, crystal damage occurs. Extreme care must be taken in designing for low power dissipation (which is a non-issue in the lumped case).
- The design is of easy implementation only for frequencies below 50 MHz. Oscillating frequencies above this are not practical.

5.5.2 Fundamental-Mode Inverter-Driven Oscillators

The symbol and the electrical equivalent of a piezoelectric crystal operating in fundamental mode (also referred to as 'parallel mode') is shown in Figure 5.23. The value C_o is the parasitic capacitance due to the crystal holder and casing. In spite of its small value, of the order of 1 pF, C_o plays a central role in the overall resonator behavior, because C_m is extremely small. In fact $C_m \ll C_o$, typically $C_m/C_o = O(1/100)$.

Some of the important crystal parameters are

- Operating frequency (ω_0): the nominal frequency of oscillation.
- Load capacitance (C_L): the capacitance that should appear in parallel to the physical terminals in order to achieve resonance at ω_0. At oscillation frequency, the load capacitance exactly resonates out the reactive part of the impedance presented by the crystal.
- Serial resonant frequency:

$$\omega_s = 2\pi f_s = \frac{1}{\sqrt{L_m C_m}} \tag{5.126}$$

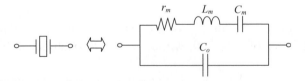

Figure 5.23   Electrical equivalent of a fundamental-mode piezoelectric crystal

- Parallel resonant frequency:

$$\omega_p = 2\pi f_p = \frac{1}{\sqrt{L_m C_m C_0/(C_m + C_0)}} \tag{5.127}$$

- Shunt capacitance (C_o): the parasitic holder capacitance.
- Drive level (P_o): the maximal allowable power dissipation on crystal.
- Quality factor:

$$Q = \frac{\omega_0 L_m}{r_m} \tag{5.128}$$

Representative values are:

$$f_0 = 8.0 \ MHz, \quad Q = 20{,}000, \quad r_m = 30 \ \Omega, \quad C_m = 0.018 \ pF,$$

$$L_m = 22 \ mH, \quad C_o = 4.5 \ pF, \quad C_L = 13 \ pF.$$

All we need in order to carry out the oscillator design, however, is the knowledge of f_s, f_p, C_o and Q. Note that (5.126) and (5.127) imply

$$f_s < f_0 < f_p \tag{5.129}$$

All these required values are easily measured in the laboratory as we explain later, and once they are known, all the motional values L_m, C_m and r_m can also be computed from the expressions above.

Since the combination of L_m, C_m, C_o, and C_L must be resonant at f_0 in the topology of Figure 5.5, which includes the circuit capacitances C_I and C_V, then it is evident that the combination of the crystal elements must present inductive reactance at this frequency. It follows that, at $\omega_0 = 2\pi f_0$, the crystal itself can be seen as an inductive reactance $X_R(\omega_0)$ in series with an effective serial resistance $r_e(\omega_0)$, as shown in Figure 5.24.

Note that both X_R and r_e depend on frequency. In particular (the proof is left as an exercise to the reader)

$$r_e(\omega) = \frac{1}{2QC_0} \frac{\omega_p - \omega_s}{(\omega - \omega_p)^2}$$

$$X_R(\omega) = j\frac{1}{\omega C_0} \frac{\omega - \omega_s}{\omega_p - \omega} \tag{5.130}$$

The topology of Figure 5.24 leads to the feedback network configuration of Figure 5.5, therefore, it is not surprising that the standard configuration of CMOS

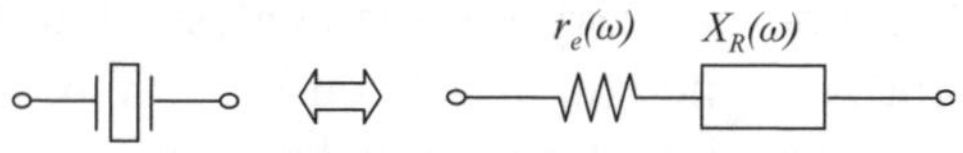

Figure 5.24 Serial equivalent of a crystal resonator at $\omega = \omega_0$

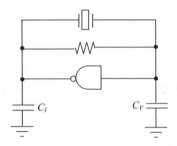

Figure 5.25   CMOS inverter-gate driven crystal oscillator topology

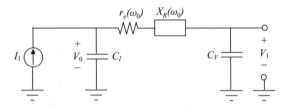

Figure 5.26   Electrical equivalent of Figure 5.25

inverter-gate driven crystal oscillators looks as shown in Figure 5.25 (bias not shown).

The resistor has the purpose of keeping the logic gate in the linear region, and it has large values of the order of >1 MΩ. The electrical circuit corresponding to Figure 5.25 at $\omega = \omega_0$ is shown in Figure 5.26. The large resistor has no effect and is omitted.

We anticipate the following (which the reader can easily infer from the analysis done with the bipolar transistor):

- The value of the serial combination of C_I and C_V must be equal to the required load capacitance C_L

$$C_L = \frac{C_I C_V}{C_I + C_V} \tag{5.131}$$

- Just as seen in (5.73), the amplitude of the oscillations at gate input will be maximal if we chose $C_I = C_V = 2C_L$. This is also the best choice as far as design stability and reliability are concerned.
- The output may be taken either from C_I or C_V, where the latter choice yields a 'cleaner' sinusoidal wave due to the filtering action of the feedback network.

All we have to do in order to carry out the analysis is to find the large-signal fundamental component I_1 of the output gate current (at $\omega = \omega_0$). The rest of the analysis is identical to the one done for the bipolar transistor (most other nonlinear devices follow the same lines).

The logic gate has a limited current sink/source capability. Denote by I_0 the maximal value of current sink/source that the gate output is capable to provide. Since the voltage gain of the inverter is extremely large, then, we can see the gate output as a current source $i(t)$ of fixed amplitude I_0, whose polarity changes according to the polarity of the input voltage $V_1 \cos(\omega t)$: if the input voltage is positive the output is in sink mode, else is in source mode. Formally, referring to Figure 5.26, we may write, at $\omega = \omega_0$

$$i(t) = -I_0 \, sign(V_1 \, cos(\omega_0 t)) \tag{5.132}$$

We may now expand $i(t)$ into a Fourier series obtaining the fundamental current I_1

$$i(t) = -\frac{4}{\pi} I_0 \cos(\omega_0 t) + \frac{4}{3\pi} I_0 \cos(3\,\omega_0 t) + \cdots, \Rightarrow I_1 = -\frac{4}{\pi} I_0 \tag{5.133}$$

Assuming that the feedback network filters out the higher harmonics, for $\omega = \omega_0$ we obtain from (5.5)

$$V_1 = -\frac{I_1}{\omega_0^2 C_I C_V r_e(\omega_0)}, \quad I_1 = -\frac{4}{\pi} I_0 \tag{5.134}$$

If we set $C_I = C_V = 2C_L$ then, in Figure 5.26 $V_0 = -V_1$, and using (5.134), the total power delivered to the crystal resonator (to the lossy feedback network) is

$$P_o = \frac{1}{2} V_0 I_1 = -\frac{1}{2} V_1 I_1 = \frac{1}{2} \frac{I_1^2}{\omega_0^2 C_I^2 r_e(\omega_0)} = \frac{8 I_0^2}{\pi^2 \, \omega_0^2 C_I^2 r_e(\omega_0)} \tag{5.135}$$

We recall that P_o must be carefully controlled to avoid crystal damage.

If P_o comes out too large from (5.135), then we must reduce I_0. This can be done by placing a limiting resistor R_L between the gate output and the feedback network, which limits the sink/source current, as shown in Figure 5.27.

If the inverter is operated with a supply voltage V_{CC}, and oscillates between $+V_{CC}$ and ground, then the average (DC) output voltage is $V_{CC}/2$. Therefore, with R_L in place, the Norton's equivalent is a source/sink current $I_0' = V_{CC}/2R_L$ in

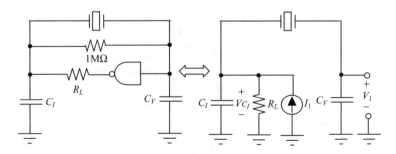

Figure 5.27 Limiting crystal dissipation

parallel with the resistor R_L. Of course, the limiting resistor will reflect in series to the crystal causing some degradation of its Q value. However, in most clock applications it does not matter, since the critical issues are frequency accuracy and stability rather than phase noise.

Computing R_L is a straightforward task. First we disregard the >1 MΩ resistor. Then we use the serial-to-parallel transformation (5.69) to compute the equivalent resistance $R_e(\omega_0)$ in parallel to C_I due to the presence of $r_e(\omega_0)$

$$R_e \approx Q_e^2 r_e, \quad Q_e = \frac{1}{\omega_0 C_I r_e}, \quad Q_e^2 \gg 1 \tag{5.136}$$

Since the DC voltage on the capacitor C_I is $V_{CC}/2$, as long as it does not exceed the maximal gate capability value I_0, the source/sink current is

$$I_0' = \frac{V_{CC}}{2R_L}, \quad I_0' \leq I_0 \tag{5.137}$$

and from (5.133), the peak fundamental current is

$$I_1 = -\frac{4}{\pi} I_0' = -\frac{2}{\pi} \frac{V_{CC}}{R_L} \tag{5.138}$$

The peak voltage on C_I is

$$V_{C_I} = I_1 \frac{R_L R_e}{R_L + R_e} = -\frac{2}{\pi} V_{CC} \frac{R_e}{R_L + R_e} \tag{5.139}$$

and the power P_o dissipated on the crystal is

$$P_o = \frac{1}{2} \frac{V_{C_I}^2}{R_e} = \frac{2}{\pi^2} V_{CC}^2 \frac{R_e}{(R_L + R_e)^2} = \frac{2}{\pi^2} V_{CC}^2 \frac{R_e}{(R_L + R_e)^2} \tag{5.140}$$

From (5.140) we compute

$$R_L = \sqrt{\frac{2R_e}{P_o} \frac{V_{CC}}{\pi}} - R_e, \quad I_0' \leq I_0 \tag{5.141}$$

Note that due to the limitation $I_0' \leq I_0$, (5.141) is valid only for

$$R_L \geq \frac{V_{CC}}{2I_0} \tag{5.142}$$

If $0 < R_L < V_{CC}/2I_0$ the gate will act as a fixed current source/sink during parts of the cycle, and will be in voltage saturation for the rest, yielding a result dependent on component variations (not a very good choice for a stable design).

As pointed out before, the oscillating frequency will be such that the load capacitance C_L 'resonates out' the serial reactance X_R, thus the requirement is

$$\frac{1}{j\omega_0 C_L} + X_R(\omega_0) = 0 \tag{5.143}$$

From (5.130) it follows that, at resonance

$$\frac{1}{\omega_0 C_0} \frac{\omega_0 - \omega_s}{\omega_p - \omega_0} = \frac{1}{\omega_0 C_L} \tag{5.144}$$

Recalling that $C_m/C_o \ll 1$, and in view of (5.126) we may approximate (5.127) as follows

$$\omega_p = \frac{1}{\sqrt{L_m C_m C_0/(C_m + C_0)}} = \frac{1}{\sqrt{L_m C_m}\sqrt{1/(1 + C_m/C_0)}} \approx \omega_s \left(1 + \frac{C_m}{2C_0}\right) \tag{5.145}$$

thus $\omega_p > \omega_s$ always. Substituting (5.145) into (5.144) yields

$$\omega_0 \approx \omega_s \left[1 + \frac{C_m}{2(C_L + C_0)}\right] \tag{5.146}$$

So the oscillating frequency depends on the loading capacitance, and

$$\lim_{C_L \to 0} \omega_0 = \omega_p, \quad \lim_{C_L \to \infty} \omega_0 = \omega_s \tag{5.147}$$

It may be objected that the difference between ω_p and ω_s is small, actually of the order of 100 PPM, and therefore the value of C_L is not that critical for clock applications. Unfortunately this is not the case. Using a value of C_L other than specified, not merely shifts the frequency, but may result in serious malfunctioning. To see this consider $r_e(\omega)$ in (5.130). If ω_0 is the oscillating frequency corresponding to the nominal value of C_L, it is easy to verify that

$$\lim_{\omega \to \omega_p} r_e(\omega) = \infty, \quad \lim_{\omega \to \omega_s} r_e(\omega) = r_s < r_e(\omega_0) \tag{5.148}$$

Then, in view of (5.147) the oscillation voltage V_1 given in (5.134)

- Decreases when C_L is smaller than specified which ultimately results in erratic start up or no oscillations at all.
- Increases when C_L is larger than specified, ultimately resulting in over-dissipation and damage of the crystal, as shown by (5.135).

Let us see a numerical example

Example: The inverter gate works with supply voltage $V_{CC} = 3$ V and has a source/sink current capability of $I_0 = 2$ mA. The measured data for a crystal to be operated in an oscillator at frequency $f_0 = 2.600000$ MHz with allowable dissipation $P_o \leq 100$ μW is

$$f_s = 2.598989 \; MHz, \quad f_p = 2.605480 \; MHz, \quad Q = 40,000, \quad C_0 = 3 \; pF$$

Using (5.130) we compute

$$r_e(\omega_0) \approx 140 \ \Omega, \quad X_R(\omega_0) \approx j3764 \ \Omega$$

Since C_L must resonate out X_R, then

$$\frac{1}{j\omega_0 C_L} + X_R(\omega_0) = 0$$

and therefore

$$C_L = \frac{1}{\omega_0 |X_R(\omega_0)|} = 16.2 \ pF \Rightarrow C_I = C_V = 2C_L \approx 32 \ pF$$

Now, with a safe requirement of $P_o \leq 50 \ \mu W$, we get from (5.135) the bound $I_0 \leq 49 \ \mu A$. This bound is much smaller than the sink/source capability of the gate, therefore a limiting resistor is required. We compute Q_e and R_e from (5.136)

$$Q_e = \frac{1}{\omega_0 C_I r_e} \approx 13.7 \Rightarrow Q_e^2 \approx 188 \gg 1 \Rightarrow R_e \approx Q_e^2 r_e \approx 26,320 \ \Omega$$

and then with $P_o = 5 \cdot 10^{-5} \ W$, we get from (5.141)

$$R_L \approx 4.7 \ k\Omega$$

With these values of R_L and R_e we get from (5.137) and (5.139)

$$I_0' \approx 0.3 \ mA, \quad V_{C_I} \approx 1.6 \ V$$

and the total equivalent resistance R_T in parallel to C_I is

$$R_T = \frac{R_L R_e}{R_L + R_e} \approx 4 \ k\Omega$$

Since the quality factor of the resonating network is inversely proportional to the energy dissipation, then the Q of the crystal resonator is reduced as follows

$$Q_{loaded} = Q\frac{R_T}{R_e} = 40,000\frac{4,000}{26,320} \approx 6,100$$

which is usually satisfactory for proper clock operation.

5.6 Measurement of Crystal Parameters

Due to the great diversity of possible implementations, crystal information provided by manufacturers is often insufficient for proper oscillator design. Fortunately, however, a precise measurement of crystal parameters can be carried out

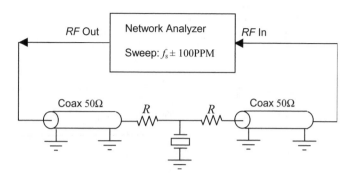

Figure 5.28 Crystal measurement setup

with the use of standard lab equipment. In particular, what we need to find out, are the values of r_m, L_m, C_m and C_o in Figure 5.23.

The measurement becomes simple if we assume that (as true in most cases of interest) at the relevant frequencies, the motional resistance r_m is much smaller that the value of the reactance of the parasitic capacitor C_o. With this assumption, the ESR (equivalent serial resistance) measured across the crystal terminals at serial resonating frequency f_s, is approximately equal to r_m, and the procedure is carried out using just a network analyzer with the 50 Ω setup of Figure 5.28, where R is a resistor much larger than r_m (for the sake of crystal protection), say $R \approx 1$ kΩ.

Denote by V the RF voltage supplied by the network analyzer. At low frequencies, away from f_s, the crystal is an open circuit, and the value of R is much smaller than the impedance presented by the capacitor C_o. Therefore, the attenuation seen by the network analyzer is approximately

$$A_1 \approx 20 \, log_{10} \left(\frac{50}{2R} \right) \, [dB] \tag{5.149}$$

At the serial resonant frequency f_s, the crystal presents a small resistance approximately equal to $r_m \ll R$, therefore the attenuation is about

$$A_2 \approx 20 \, log_{10} \left(\frac{r_m \cdot 50}{R^2} \right) \, [dB] \tag{5.150}$$

The 'notch' attenuation $A_2 - A_1$ is measured as shown in Figure 5.29, and

$$A_2 - A_1 \approx 20 \, log_{10} \left(\frac{2r_m}{R} \right) \, [dB] \tag{5.151}$$

Since R is known, r_m is now directly computed from (5.151).

Now, as shown in Figure 5.30, we can measure the 3 dB 'notch' bandwidth of the serial resonant equivalent, from which the values of Q, C_m and L_m may be

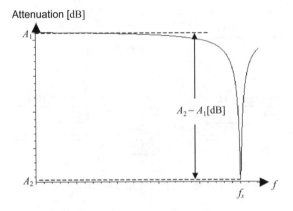

Figure 5.29   'Notch' attenuation at f_s

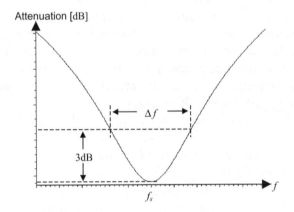

Figure 5.30   3 dB 'notch' bandwidth at f_s

readily computed from (the proof is left to the reader)

$$\frac{f_s}{\Delta f} = Q = \frac{1}{2\pi f_s r_m C_m} = \frac{2\pi f_s L_m}{r_m} \tag{5.152}$$

Since C_o is a physical capacitor, its value is measured directly, with conventional means (a network analyzer or a capacitance meter), at a frequency lower than f_s (say, below $f_s/2$) where the crystal is an open circuit, but still high enough to enable an easy measurement of values of the order of 1 pF.

5.7 Measurement of Oscillators

The measurement of oscillators is complex and tricky, and often is carried out by means of special dedicated setups. Here we bring what we believe is a simple, accurate and practical approach using standard lab equipment. Before we proceed, however, we need to gain some more insight into the correspondence between SSB phase noise and narrowband FM modulation.

5.7.1 Phase Noise as Narrowband FM

The name 'narrowband FM' refers to a particular case of FM modulation. We recall that an FM-modulated RF signal can be represented in the form (Clarke, 1978; Millman-Halkias, 1972)

$$S(t) = A \cos(\phi(t)) = A \cos\left[\omega_0 t + \Delta\omega \int_0^t s(\tau) \, d\tau\right], \quad \|s(\tau)\|_\infty \le 1 \quad (5.153)$$

where A is a positive constant, $\phi(t)$ is the instantaneous phase, ω_0 denotes the angular frequency of the carrier, $\Delta\omega$ is called the 'peak deviation' of the angular frequency, and $s(\tau)$ is a modulating signal whose absolute amplitude does not exceed unity.

For ease of understanding, we consider the case where

$$s(t) = \cos(\omega_m t), \quad |\omega_m| \ll |\omega_0| \quad (5.154)$$

Using (5.154), Equation (5.153) becomes

$$S(t) = A \cos[\omega_0 t + \beta \sin(\omega_m t)], \quad \beta \equiv \left|\frac{\Delta\omega}{\omega_m}\right| \quad (5.155)$$

where β is referred to as the 'modulation index'.

$S(t)$ is referred to as a 'narrowband FM' signal if $\beta \ll 1$. In this case, we may approximate (the proof is left to the reader)

$$S(t) \approx A[\cos(\omega_0 t) - \beta \sin(\omega_m t) \sin(\omega_0 t)]$$
$$= A\left[\cos(\omega_0 t) + \frac{\beta}{2} \cos((\omega_0 + \omega_m)t) - \frac{\beta}{2} \cos((\omega_0 - \omega_m)t)\right] \quad (5.156)$$

Thus $S(t)$ in (5.156) consists of the carrier and one pair of (small) sidebands, each of amplitude $\beta/2 \ll 1$ relative to the carrier amplitude.

We observe that a narrowband FM signal differs from an AM signal because the sidebands are out of phase by 180° rather than in-phase.

Since oscillators have constant amplitude, any low-level spread-out of their spectral picture away from the center frequency ω_0, may be thought as being produced by narrowband FM modulation. In particular, it is customary to relate the total RF power contained within a 1 Hz bandwidth around the RF frequencies $\omega = \omega_0 \pm \omega_m$ to an equivalent sinusoidal signal of frequency ω_m modulating the oscillator carrier in narrowband FM mode with a modulation index β.

Specifically, with reference to (5.10) and (5.156) in a 1 Hz bandwidth

$$L(\omega - \omega_0) = L(\omega_m) \approx 20 \log_{10}(\beta/2) \quad [dBc/Hz] \quad (5.157)$$

which produces a discrete representation of the spectral picture with a 1 Hz resolution.

NOTE: (5.157) shows that there is a one-to-one correspondence between the power density of the phase noise at any given offset from the carrier and the sideband power of an equivalent narrowband FM modulated signal at the same offset. Therefore it is customary to denote the phase noise by the name 'single sideband (SSB) noise'.

Example: We compute the equivalent SSB noise of a 100 MHz VCO due to the presence of a sinusoidal low-frequency interferer $s(t)$ on the steering line (recall: 'steering line' $\equiv$ the frequency-control port of the VCO).

The sensitivity of the steering line is $K_V = 20$ MHz/V, and the characteristics of the interferer are

$$s(t) = A \cos 2\pi f_m t, \quad A = 0.1 \ mV, \quad f_m = 500 \ kHz$$

which yields $\Delta f = K_V A = 2 \ kHz$, $\beta = \Delta f / f_m = 4 \cdot 10^{-3}$, and finally

$$L(\pm 500 \ kHz) = 20 \log_{10}(\beta/2) \approx -54 \ dBc/Hz$$

5.7.2 Single Sideband (SSB) Noise

The measurement of SSB noise is accomplished using the setup described in Figure 5.31. The way of operation of the setup is not trivial. Let us first see what happens in the mixer. The mixer multiplies a 'noisy' VCO signal $S(t)$, with zero phase offset and angular frequency ω_0, by a carrier signal $S_c(t)$ near the same center frequency, narrowband-FM-modulated by a sinusoid at baseband frequency ω_m and with a fixed relative average phase offset ϕ and unit amplitude

$$S(t) = A \cos[\omega_0 t + n(t)], \quad \|n(t)\|_\infty \ll 1, \quad E[n(t)] = 0$$
$$S_c(t) = \cos[\omega_0(1 + \varepsilon)t + \phi + \beta \sin(\omega_m t)], \quad \beta, \varepsilon \ll 1 \qquad (5.158)$$

The noise $n(t)$ in (5.158) is a sample function from a zero-mean ergodic random process as described in section 5.3.

The filter $LPF1$ has low-pass characteristics with cutoff at very low frequency, of the order of 10 Hz or less. Assuming that the closed-loop system of Figure 5.31 ultimately reaches a steady-state condition, after the signal generator has been adjusted to an average open-loop center frequency $\omega_0(1 + \varepsilon)$, we show that the output of $LPF1$ must stabilize to a fixed DC voltage $|V_{LP1}| \approx 0$ corresponding to a phase offset $|\phi| \approx \pi/2$, and the signal generator locks up to the frequency ω_0. Indeed, it follows from (5.158) that the low-pass output of the mixer is

$$V_{LP1} = S(t) \cdot S_c(t)|_{LP1} \approx \frac{A}{2} \cos \phi \qquad (5.159)$$

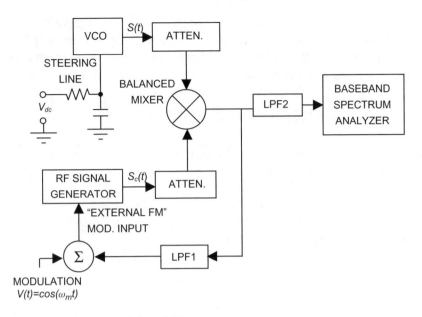

Figure 5.31   SSB noise – measurement setup

Assume that the sensitivity at the 'external FM' modulation port of the generator is K_F [MHz/V]. The DC voltage V_{LP1} out of $LPF1$ causes the generator frequency to change until lock up occurs when $K_F V_{LP1} - \varepsilon\omega_0 = 0$, namely

$$V_{LP1} = \frac{\omega_0}{K_F}\varepsilon \qquad (5.160)$$

If the initial adjustment of the signal generator is close to ω_0, and the modulation sensitivity is high, then $|V_{LP1}|$ is small, and $|\phi| \approx \pi/2$. Note that (5.159) implies that ϕ must be close to either $\pi/2$ or $-\pi/2$, depending of the polarity of the feedback.

We assume without loss of generality that $\phi = \pi/2$, then, under steady-state conditions, the lower equation in (5.158) becomes

$$S_c(t) = -\sin[\omega_0 t + \beta \sin(\omega_m t)], \quad \beta \ll 1 \qquad (5.161)$$

The filter $LPF2$ exhibits DC blocking and low-pass characteristics with cutoff frequency slightly above the largest offset value ω_m of interest.

With the help of (5.161) and the upper equation in (5.158), the signal V_{LP2} at the output of $LPF2$ takes the form

$$V_{LP2} = S(t) \cdot S_c(t)|_{LP2} = \frac{A}{2}\sin[n(t) - \beta \sin(\omega_m t)] \approx \frac{A}{2}n(t) - \frac{A\beta}{2}\sin(\omega_m t) \qquad (5.162)$$

Now, it follows from section 5.3 that since $n(t)$ is a random process it has a spectral density $R(\omega)$ [watt/Hz].

Assume that $R(\omega)$ can be considered constant over the bandwidth B of the IF filter of the spectrum analyzer. With the modulation turned off, namely $\beta = 0$, the noise power $P_n(\omega_m)$ detected by the analyzer within an IF bandwidth centered at frequency offset ω_m is proportional to the carrier power and follows from (5.162)

$$P_n(\omega_m) \approx R(\omega_m)BP_0, \quad P_0 = \left(\frac{A}{2}\right)^2 \tag{5.163}$$

where P_0 is the power of the unmodulated carrier. So we may think of $P_n(\omega_m)$ as being the sideband power of a discrete narrowband FM signal at frequency ω_m caused by the noise modulating the carrier.

With the modulation turned on, if the discrete sideband power $P_c(\omega_m)$ at frequency ω_m due to the modulation of the carrier with and index $\beta \neq 0$ is much larger than $P_n(\omega_{m,})$, again by (5.162)

$$P_c(\omega_m) \approx \frac{1}{2}\beta^2 P_0, \quad P_0 = \left(\frac{A}{2}\right)^2 \tag{5.164}$$

Figure 5.32 shows ratio Δ_m between $P_c(\omega_m)$ and $P_n(\omega_m)$ measured in dB on the spectrum analyzer, for a given IF bandwidth B. The modulation index β is adjusted to satisfy $P_c(\omega_m) \gg P_n(\omega_{m,})$.

From (5.163) and (6.164) we may compute $R(\omega_m)$, which is the noise power per unit bandwidth relative to carrier power at frequency ω_m

$$\Delta_m = 10\log_{10}[P_c(\omega_m)/P_n(\omega_m)] = 10\log_{10}[\beta^2/2R(\omega_m)B]$$
$$= 20\log_{10}[\beta] - 3 - 10\log_{10}[R(\omega_m)] - 10\log_{10}[B] \tag{5.165}$$

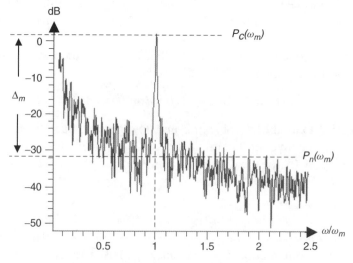

Figure 5.32   SSB noise computation by narrowband FM comparison

From which we finally obtain

$$L(\omega_m) = 10\,log_{10}[R(\omega_m)] = 20\,log_{10}[\beta] - \{\Delta_m + 3 + 10\,log_{10}[B]\} \quad \left[\frac{dBc}{Hz}\right]$$
(5.166)

Example: Assume that in Figure 5.32

- $\omega_m/2\pi = 25$ kHz, the offset from carrier.
- $\Delta\omega/2\pi = 100$ Hz, the peak deviation of the modulation.
- $B = 10$ kHz, the IF bandwidth of the spectrum analyzer.
- $\Delta_m \approx 34$ dB, the ratio $P_c(\omega_m)/P_n(\omega_m,)$ in dB.

Then the modulation index is $\beta = 100/25,000 = 0.004$. Using (5.166)

$$L(\omega_m) = 20\,log_{10}(0.004) - \{34 + 3 + 10\,log_{10}(10,000)\}$$
$$\approx -48 - \{34 + 3 + 40\} \approx -125\ dBc/Hz\,@25\ kHz \quad (5.167)$$

NOTE: In order to obtain an accurate measurement, we must be aware of the following:

- The signal generator must have an SSB phase noise at least 10 dB better than the VCO at the relevant frequencies, otherwise degradation will occur. Using an RF generator with a phase noise worse than the VCO target spec is a common pitfall for engineers, and ends up in measuring the generator values, so no matter how good the VCO is, it will look out of specification. The same holds for pointwise generator spurs, yielding wrong readings at the corresponding frequencies.
- In order to lock-up to VCO frequency, the signal generator must have 'external FM' modulation capability down to DC. However, the carrier modulation at frequency ω_m may be set by using internal FM capability as well. The modulation index β will be increased until the discrete sideband at ω_m is well above the VCO noise level.
- The attenuators (≥ 10 dB attenuation each) in the setup of Figure 5.31 are an absolute requirement. This is because the RF injection from the signal generator driving the balanced mixer must have a high power level. If the generator leaks, mutual injection locking may occur (Chapter 6), which will make even a poor VCO noise look as good as the generator noise. To improve isolation a double-balanced mixer is preferable.
- Since the mixer acts as a phase detector, it should be selected for improved linearity and gain of phase-to-voltage conversion.
- If the VCO is prone to AM noise, a limiter should be added at its output to prevent AM fluctuations from reaching the spectrum analyzer.
- $LPF2$ must be wide enough to cover the whole required range of ω_m.

- *LPF*1 should be very narrow, down to few Hz, and the overall loop behavior may be computed with the help of the basic PLL theory of Chapter 4. In this case the loop divider is equal to one, and the phase detector gain is the mixer gain.
- The VCO is set at the desired center frequency by applying a suitable external DC voltage to the steering line. Since the steering sensitivity is usually very high, of the order of many MHz/V, we must provide very good filtering, otherwise noise coming through the DC line will modulate the VCO producing high SSB noise readings. The resistor–capacitor arrangement in Figure 5.31 has exactly this purpose.
- The 3 dB corner of the RC combination is usually of the order of a fraction of Hz. The capacitor must be large, otherwise the thermal noise generated within the resistor will modulate the VCO altering the measurement. In fact, using a large resistor on the steering line is a common design mistake, and often becomes a limiting factor to the overall VCO noise performance.

5.7.3 Residual FM

Residual FM is an approximated, yet extremely simple and useful measurement method to estimate the SSB noise of an oscillator, by using only a modulation analyzer working in FM mode.

The term 'residual FM' is a legacy from analog FM receivers. It was noted that, when connecting a signal generator delivering a clean 'silent' (unmodulated) and strong carrier to the antenna port of a FM receivers, we get a background 'hiss' noise in lieu of the silence that could have been expected, no matter how strong and clean the input signal is made. This effect is connected to the phase noise of the receiver LO. As we pointed out before, the Rx mixer causes the input carrier to acquire the very same phase noise of the LO. This phase noise, upon reaching the FM demodulator, is converted into a baseband noise (the 'hiss').

The bright side of it is that, by measuring the baseband noise obtained by bandpass-filtering the demodulator output, we can get a fair estimate of the SSB noise at any distance from the carrier (above noise floor).

Consider the constant-amplitude (normalized) RF output of an oscillator, which includes low-level random phase noise as in (5.158), and thus $n(t)$ is a random variable

$$S(t) = cos[\omega_0 t + n(t)], \quad \|n(t)\|_\infty \ll 1, \quad E[n(t)] = 0 \tag{5.168}$$

When the above signal undergoes FM detection, the FM discriminator performs first the time differentiation of $S(t)$, then AM detection, and finally DC rejection (Schwartz, 1990). The differentiation action yields

$$\frac{d}{dt}S(t) = \left[\omega_0 + \frac{d}{dt}n(t)\right] sin[\omega_0 t + n(t)] \tag{5.169}$$

After AM detection and DC rejection the detected baseband signal $r(t)$ is

$$r(t) = \frac{d}{dt} n(t) \tag{5.170}$$

As in the previous section, denote by $R(\omega)$ the spectral density of the random variable $n(t)$. Note that $n(t)$ is in radians [rad] and therefore $R(\omega)$ has dimension [rad^2/Hz].

We recall that time differentiating a signal $g(t)$ is equivalent, in the Fourier domain, to filtering it with a linear filter $H(\omega) = j\omega$ (Schwartz, 1990)

$$F\left[\frac{d}{dt} g(t)\right] = j\omega F[g(t)] \tag{5.171}$$

We also recall from (5.20) that when a random process passes through a linear filter, the spectral density at the filter output is equal to the input spectral density multiplied by $|H(\omega)|^2$

$$R_{out}(\omega) = |H(\omega)|^2 R(\omega) = \omega^2 R(\omega) \tag{5.172}$$

where $R_{out}(\omega)$ denotes the spectral density out of the FM detector, and we recall from (5.166) that with $\omega_m = \omega - \omega_0$ denoting the offset frequency from the RF carrier

$$10 \log_{10}[R(\omega_m)] = L(\omega_m), \quad \omega_m = \omega - \omega_0 \tag{5.173}$$

Thus, if we know $R_{out}(\omega_m)$ for some ω_m we may compute $L(\omega_m)$ from (5.172) and (7.173). At this point we note that the value $\omega_m^2 R(\omega_m)$ is constant. This is easily seen by looking at Leeson's equation (5.10).

Denoting by B the bandwidth of the loaded resonator, we recall that $Q_L = \omega_0/B$, and therefore

$$\left(2Q_L \frac{\omega_m}{\omega_0}\right)^2 = \left(2\frac{\omega_m}{B}\right)^2 \ll 1, \quad Q_L = \frac{\omega_0}{B} \tag{5.174}$$

That is for ω_m much smaller that the resonator bandwidth (which is usually the practical case), we may approximate, using (5.10)

$$R(\omega_m) \approx \frac{kTFB^2}{8P_s} \frac{1}{\omega_m^2} \tag{5.175}$$

and therefore

$$\omega_m^2 R(\omega_m) \approx \frac{kTFB^2}{8P_s} = \text{constant} \tag{5.176}$$

Integrating $R_{out}(\omega)$ in [rad^2/Hz] within a bandpass filter above the flicker corner, say between B_1 and B_2, $B_1 < B_2$, the measured output baseband power P_{bb} is

$$P_{bb} = \int_{B_1}^{B_2} f_m^2 R(2\pi f_m) df_m = (B_2 - B_1) f_m^2 R(2\pi f_m) \tag{5.177}$$

If the modulation analyzer reads a rms deviation Δf_{rms} [Hz], then we can take the approximation

$$\Delta f_{rms}^2 \approx P_{bb} \tag{5.178}$$

and get from (5.177)

$$R(2\pi f_m) = \frac{\Delta f_{rms}^2}{(B_2 - B_1) f_m^2} \tag{5.179}$$

and then again compute $L(\omega_m)$ with (5.173).

Note that, by virtue of Leeson's equation, although the measurement is done within $B_2 - B_1$, we may extrapolate the result as long as (5.174) holds.

Example: The post-detection filter of the modulation analyzer is set for the range 30 Hz–3 kHz. When testing a synthesizer, the reading on the analyzer shows a residual FM deviation of 2 Hz rms. We want to compute the phase noise at 25 kHz offset from the carrier.

Using (5.179) we compute

$$R(2\pi 25\ kHz) = \frac{4}{2950 \cdot (25 \cdot 10^3)^2} \approx 2.17 \cdot 10^{-12} \tag{5.180}$$

and therefore (5.173) yields

$$L(2\pi 25\ kHz) = 10\ log_{10}(2.17 \cdot 10^{-12}) \approx -116.6\ dBc/Hz \tag{5.181}$$

5.7.4 Frequency Pushing/Pulling

The term 'frequency pushing' refers to the tendency of the VCO to shift in frequency due to changes in the supply voltage. This parameter is important because, when a transmitter suddenly goes from idle to transmit mode, the abrupt change in the current drain produces a temporary voltage drop on the VCO supply. In turn, the change in supply voltage may produce a sudden shift in the VCO output frequency, which may even drive the PLL out of lock, and into a relock process. During this transient, the transmitter may cause interference to other receivers in the systems, therefore 'frequency shift during transmit transients' is usually a regulatory requirement. One way to avoid the problem is to isolate all VCO and PLL voltage regulators from the transmitter regulators.

The measurement is taken with the VCO steering line disconnected from the loop filter and connected to a fixed control voltage using an external power supply and an RC filter, just as shown in Figure 5.31. The VCO is connected to a programmable DC supply capable of switching within short times.

The term 'frequency pulling' refers to the tendency of the VCO to shift in frequency due to changes in the VCO load impedance. The load impedances are defined for various values of VSWR at all angles. This is a crucial parameter, because, when a transmitter switches from idle to transmit mode, there is an abrupt change in the input impedances of the PA and its associated drivers. Since the backward isolation of the Tx chain is finite, the load changes may reflect at the VCO output, modify the resonance conditions, resulting in a frequency shift, and consequently generating interferences. As before, the 'frequency shift during transmit transients' is affected.

One way to improve backward isolation is to alternatively add buffers and then passive attenuators in the VCO–PA path. The measurement is taken as before, disconnecting the steering line from the loop filter applying a fixed (and filtered) control voltage. The VCO output is connected to a variable load impedance that can be controlled by electrical means and switched within short times.

5.7.5 Output Flatness

VCO output flatness is the ratio in dB between the maximum and minimum carrier power measured at the VCO output on a 50 Ω load over the whole frequency and temperature ranges. Its variations are mainly due to oscillation level drop following a loaded Q drop as frequency increases.

5.8 Lumped Equivalent of Distributed Resonators

Resonating circuits operating at frequencies above 150 MHz cannot achieve good performance when built in lumped form. The reason is that, at high frequencies, lumped components exhibit unreliable behavior as well as poor quality factors. Therefore, high-frequency, high-performance resonators are constructed with the help of various distributed technologies.

Virtually all the technologies of practical use are resonant 'cuts' of electrically tuned, low-loss transmission lines, such as coaxial cables, microstrips, striplines, ceramic resonators, helical resonators etc. of electrical length close to $\lambda/4$.

Regardless of the physical resonator implementation, however, it turns out that we may always reduce the distributed resonator to a lumped equivalent. Once this lumped equivalent is computed, the VCO design may be accomplished following the very same technique used in section 5.4.2 for the lumped case.

In this section we will prove that a grounded near-quarter-wavelength transmission line is equivalent to a parallel resonant circuit, and therefore may be used to design a resonator of the type described in Figure 5.18. Regardless of the specific transmission line technology, and based only on the knowledge of the characteristic impedance Z_0 [Ω] and the attenuation constant α [Neper/m], we show how to compute the equivalent lumped R, L and C values of the resonant circuit.

Moreover, we show that, by grounding the open line end through a variable reactance, its effective electrical length and consequently the resonant frequency of the parallel equivalent may be tuned, and we compute suitable formulas for practical design. Finally, we show how a transmission line resonator fits in a Colpitts-like topology similar to the one in Figure 5.22. Transmission line theory is extensively treated in Collin (1966).

5.8.1 Resonant Low-Loss Transmission Lines

Denote by P_0 the input power fed into a perfectly terminated transmission line of characteristic impedance Z_0. Assume that the line is terminated at distance ℓ from the feeding point with a matched resistive load (a load whose resistance is equal to Z_0). Then, the power P_{Load} dissipating into the load is

$$P_{Load} = P_0 e^{-2\alpha\ell} \tag{5.182}$$

where α is referred to as the 'attenuation constant'.

It follows from the above that, if the power loss per unit length is known, then α may be readily computed from (5.182). From now on, therefore, we assume that α is known, whether by spec or by measurement. Also we assume small losses, i.e. $\alpha\ell \ll 1$.

Consider a transmission line of impedance Z_0 and length ℓ terminated with a load impedance Z_ℓ as shown in Figure 5.33.

Denoting by λ the wavelength and setting $\beta = 2\pi/\lambda$, the impedance Z_{in} seen at input of the line for $\alpha\ell \ll 1$ satisfies (Jordan, 1989):

$$\frac{Z_{in}}{Z_0} = \frac{\dfrac{Z_\ell}{Z_0} + \alpha\ell + j\left(1 + \dfrac{Z_\ell}{Z_0}\alpha\ell\right)\tan\beta\ell}{1 + \dfrac{Z_\ell}{Z_0}\alpha\ell + j\left(\dfrac{Z_\ell}{Z_0} + \alpha\ell\right)\tan\beta\ell}, \beta = \frac{2\pi}{\lambda} \tag{5.183}$$

Assume that Z_ℓ is purely reactive, namely

$$Z_\ell = jX, X \in \mathbb{R} \tag{5.184}$$

Then, after a some manipulation, (5.183) takes the form

$$\frac{Z_{in}}{Z_0} = \frac{j\left[\dfrac{X}{Z_0} + \tan\beta\ell\right] + \alpha\ell\left[1 - \dfrac{X}{Z_0}\tan\beta\ell\right]}{j\alpha\ell\left[\dfrac{X}{Z_0} + \tan\beta\ell\right] + \left[1 - \dfrac{X}{Z_0}\tan\beta\ell\right]} \tag{5.185}$$

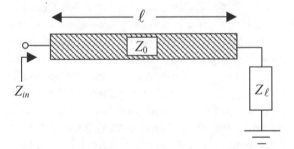

Figure 5.33 Low-loss loaded transmission line

Denote by v the wave velocity inside the transmission line, by f and λ the actual frequency and the corresponding wavelength, and by f_0 and λ_0 the frequency and the wavelength for which $\ell = \lambda_0/4$. Then, we use the well-known relations (Jordan, 1989):

$$\lambda f = v \Rightarrow \ell = \frac{\lambda_0}{4} = \frac{v}{4f_0} = \frac{\lambda}{4} \frac{f}{f_0} \tag{5.186}$$

and therefore $\beta\ell$ may be written in the form

$$\beta\ell = \frac{2\pi}{\lambda} \frac{\lambda}{4} \frac{f}{f_0} = \frac{\pi}{2} \frac{f_0 + \Delta f}{f_0} = \frac{\pi}{2} + \frac{\pi}{2} \frac{\Delta f}{f_0}, \quad \Delta f = f - f_0 \tag{5.187}$$

In the vicinity of f_0, namely if $|\Delta f/f_0| \ll 1$, we may approximate

$$\tan \beta\ell = \tan\left(\frac{\pi}{2} + \frac{\pi}{2} \frac{\Delta f}{f_0}\right) = -\cot \frac{\pi}{2} \frac{\Delta f}{f_0} \approx -\frac{1}{\dfrac{\pi}{2} \dfrac{\Delta f}{f_0}} \tag{5.188}$$

Thus, in the vicinity of f_0, $|\tan \beta\ell| \gg 1$, so, if we assume that the line is nearly short-circuited to ground and has small losses we have

$$\frac{|X|}{Z_0} \ll 1, \quad \alpha\ell \ll 1, \quad |\tan \beta\ell| \gg 1 \tag{5.189}$$

With the help of (5.189) we approximate (5.185) as follows

$$\frac{Z_{in}}{Z_0} \approx \frac{j \tan \beta\ell + \alpha\ell K}{j\alpha\ell \tan \beta\ell + K}, \quad K = \left[1 - \frac{X}{Z_0} \tan \beta\ell\right] \tag{5.190}$$

Multiplying and dividing (5.190) by the complex conjugate of the numerator and noting that

$$|\tan \beta\ell| \gg 1 \text{ and } \frac{|X|}{Z_0} \ll 1 \Rightarrow |K| \ll |\tan \beta\ell| \tag{5.191}$$

we obtain

$$\frac{Z_{in}}{Z_0} \approx \frac{\tan^2 \beta\ell + (\alpha\ell K)^2}{(j\alpha\ell \tan \beta\ell + K)(-j \tan \beta\ell + \alpha\ell K)}$$

$$\approx \frac{\tan^2 \beta\ell}{\alpha\ell[\tan^2 \beta\ell + K^2] + jK \tan \beta\ell[(\alpha\ell)^2 - 1]}$$

$$\approx \frac{1}{\alpha\ell - j\left(\dfrac{1}{\tan \beta\ell} - \dfrac{X}{Z_0}\right)} \approx \frac{1}{\alpha\ell + j\left(\dfrac{\pi}{2} \dfrac{\Delta f}{f_0} + \dfrac{X}{Z_0}\right)} \tag{5.192}$$

We see that the circuit in (5.192) is resonant at f_X when

$$\frac{\pi}{2}\frac{f_X - f_0}{f_0} + \frac{X}{Z_0} = 0 \Rightarrow f_0 = f_X + \frac{2X}{\pi Z_0} f_0 \tag{5.193}$$

Noting that (5.193) implies

$$\frac{\pi}{2}\frac{f - f_0}{f_0} + \frac{X}{Z_0} = \frac{\pi}{2}\frac{f - f_X}{f_0} = \frac{\pi}{2}\frac{f - f_X}{f_X}\left(1 - \frac{2X}{\pi Z_0}\right) \approx \frac{\pi}{2}\frac{f - f_X}{f_X} \tag{5.194}$$

we may approximate (5.192) to the form

$$Z_{in} \approx \frac{Z_0}{\alpha \ell} \cdot \frac{1}{1 + j\dfrac{\pi}{2\alpha \ell}\dfrac{f - f_X}{f_X}}, \quad f_X = \left(1 - \frac{2X}{\pi Z_0}\right) f_0 \tag{5.195}$$

where we recall that f_0 is the frequency at which the electrical length of the transmission line is exactly $\lambda_0/4$.

Equation (5.195) is the well-known impedance of a high-Q parallel resonant circuit (Clarke, 1978), which has the form

$$Z_{in} \approx \frac{R}{1 + j2Q_U\dfrac{f - f_X}{f_X}} \tag{5.196}$$

Now assuming that $X = \varepsilon Z_0$, $|\varepsilon| \ll 1$ the resonant frequency is easily computed, and we recognize from (5.195) and (5.196)

$$Q_U = \frac{\pi}{4\alpha \ell}, \quad R = \frac{Z_0}{\alpha \ell}, \quad f_X \approx f_0\left(1 - \frac{2}{\pi}\varepsilon\right), \quad \varepsilon = \frac{X}{Z_0}, \; |\varepsilon| \ll 1 \tag{5.197}$$

If the load is capacitive, then $\varepsilon < 0$, so the resonant frequency increases, and the opposite occurs if the load is inductive.

If Z_ℓ is a varactor of capacitance C_ℓ thus, provided that $1/\omega C_\ell \ll Z_\ell$

$$f_X - f_0 \approx \frac{1}{\pi^2 C_\ell Z_0} \tag{5.198}$$

The expression in (5.197) corresponds exactly to the impedance of a parallel RLC circuit resonant at f_X (Clarke, 1978) with

$$Q_U = \omega_X RC, \quad \omega_X = 2\pi f_X = \frac{1}{\sqrt{LC}} \tag{5.199}$$

and therefore we get the equivalent lumped representation

$$Q_U = \frac{\pi}{4\alpha \ell}, \quad R = \frac{4Q_U Z_0}{\pi}, \quad C = \frac{Q_U}{2\pi f_X R}, \quad L = \frac{1}{(2\pi f_X)^2 C} \tag{5.200}$$

Figure 5.34 illustrates the equivalence when Z_ℓ is capacitive.

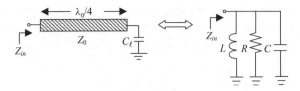

Figure 5.34   Lumped equivalent for a resonant transmission line

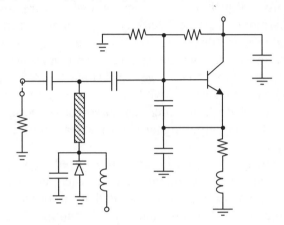

Figure 5.35   Line architecture with serial tuning ($X \neq 0$)

A possible architecture of a VCO similar to Figure 5.22 using a 'serially tuned' ($X \neq 0$) transmission line resonator is shown in Figure 5.35. In many other settings, the transmission line is 'parallel tuned', namely grounded directly ($X = 0$), and the varactor is attached at the top via a small capacitor, as shown in Figure 5.36. In this case, the varactor appears in parallel to the equivalent lumped

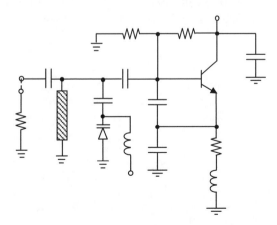

Figure 5.36   Line architecture with parallel tuning ($X = 0$)

capacitor of the parallel resonant circuit, and this is why its capacitance must be reduced using a serial capacitor as in Figure 5.36. Note the capacitor in parallel to the varactor in Figure 5.35, whose function is both to reduce $|X|$ and to control the tuning range.

5.9 Harmonic Oscillators

As pointed out before, Equation (5.56) implies that, for large oscillations (when $x \to \infty$), all the harmonics in the collector current of a bipolar transistor approach the same amplitude value (this is generally true also in clamping-mode JFET devices, and more in general, whenever large oscillations lead to small device conduction angles). It follows that if we can extract the Nth harmonic collector currents instead of the fundamental, while not impairing the original design, we obtain a 'harmonic oscillator', meaning that, although we design the oscillator to work at frequency f_0, the output signal is an integer multiple of the oscillating frequency, namely $N \times f_0$.

This can be extremely useful, for instance, if we need a crystal oscillator above 50 MHz, or if the required VCO tuning range is wide, and makes it difficult to design the resonator. The only price to pay is that the phase noise increases by $20 \log(N)$ [dB] since the FM modulation index is multiplied accordingly.

With reference to Figure 5.13, we show in Figure 5.37 that, since an ideal device behaves as a current source, if we may make the harmonic-extracting network appear as an impedance Z_H connected in series with the collector, this will not affect the computations involved in the oscillator design. In practice, since the device is not ideal but has parasitic internal feedback connections, the above statement will fail to hold if a substantial voltage develops across Z_H.

However, if Z_H is an effective short circuit at resonating frequency, and develops only a small voltage across it just at the Nth harmonic frequency so that instantaneous collector saturation does not occur, then the oscillator will work exactly as in the fundamental-mode design. For instance, the above can be

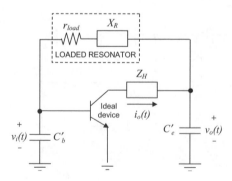

Figure 5.37   The effect of an harmonic-extracting network

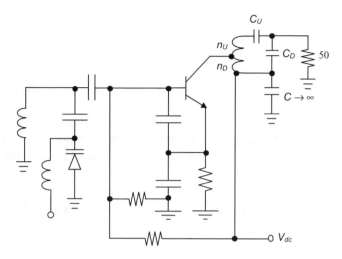

Figure 5.38   Harmonic oscillator

achieved if Z_H is the input impedance of a parallel RLC impedance transformer resonating at $N \times f_0$.

A possible implementation is shown in Figure 5.38. Here the output circuit consists of the combination of an autotransformer and a capacitive divider resonant at the Nth harmonic of the oscillating frequency.

The loading points of the output circuit should be chosen so that the loaded Q at output frequency is as high as possible while still covering the required output range, and the output-to-input transformation ratio must be such that the reflected impedance Z_H at the collector is low enough at all frequencies to prevent instantaneous saturation with the peak current computed in (5.62). If the amplitude of the oscillations is high enough, the peak harmonic current will approach the asymptotic value of $2I_{dc}$ according to (5.63).

The configuration of Figure 5.38 has a side benefit: the output does not load the resonant network of the oscillator, thus, the resonator Q is not reduced by the presence of the load, which somewhat compensates for the increased phase noise due to frequency multiplication.

5.10 A Unified Approach to Oscillator Design

Looking back to the two cases of bipolar VCO and inverter-driven crystal oscillators, we see that their design procedure is identical, except for the fact that their peak fundamental RF current I_1 is different and must be computed ad-hoc.

We stress again that the key for their very stable behavior and repeatable performance lies in the fact that the value of I_1 reaches an ultimate asymptotic value when the input peak RF drive V grows towards infinity. It follows that as $V \to \infty$, the output current source becomes independent from the input drive level V, which implies that input and output are effectively disconnected.

We conclude that the very same design procedure holds for any device that satisfies

$$\lim_{V \to \infty} I_1(V) = constant < \infty \tag{5.201}$$

and we may expect a stable and repeatable performance for all devices of this kind.

In other cases, such as JFET-based oscillators, Equation (5.201) does not hold naturally, but, including a proper V-dependent feedback in their DC bias, usually we may approximately satisfy it ($I_1(V) \approx$ constant at the oscillation level V). If (5.201) is not approximately satisfied, or if the oscillation level is designed too low so the input–output dependency is not broken, we should expect poor repeatability unless an external AGC (automatic gain control) circuit is added.

6

Parasitic and Nonlinear Phenomena

RF equipment is prone to a great deal of parasitic and nonlinear phenomena. Sometimes we can intentionally introduce them and turn their effect to an advantage, but more often they 'sneak-in' unwanted and may impair or even paralyze system operation.

Once such phenomena show up *a posteriori* (mostly as a result of overlooking their severity during the design phase), apart from a substantial design/layout modification, often very little can be done to get rid of them, and in many cases an irreversible disaster results. Several preventive actions can be taken, most of them mainly related to a proper choice of transceiver architecture (Chapter 1).

It is vital for the design engineer to be aware of parasitic and nonlinear phenomena and learn how to control such events.

6.1 Parasitic Effects in Oscillators and Synthesizers

During ongoing transceiver operation, the various oscillators within the equipment are affected by several types of disturbances:

- Direct radiation from the RF power amplifier (PA) onto a VCO resonator triggers a phenomenon known as 'oscillator pulling', and 'remodulation'.
- Intermittent PA activation, such as in modern GSM systems, results in sudden changes of ground currents and generates spurs near the transmitted carrier.
- Sharp changes in oscillator load impedance, due to sharp changes in the input impedance of the VCO buffer during its activation, tune out the oscillator resonator producing the effect known as 'frequency shift upon keying'.
- Mechanical vibrations of metallic covers in physical proximity to the resonator of a VCO may alter the resonating frequency by capacitive coupling, producing a phenomenon known as 'microphonics'. By tapping on the radio, with the transmitter in unmodulated (CW) mode and attached to an FM modulation

analyzer, we will be able to hear the noise as tapping on a microphone. The same effect may result from vibrating components in the resonator circuit, including multilayer chip capacitors or inductors. When microphonics occurs in the receiver LO, it introduces parasitic FM modulation, which leads to dramatic sensitivity and selectivity performance degradation (Chapter 2). Microphonics is a major system killer in mobile environments, and must be accurately taken into account in the design phase, since once it shows up, very little can be done, even with substantial electrical and mechanical redesign.

- Combinations of the above phenomena occur when the modulation type implies carrier envelope variations influencing PA current drain, or when some non-related on-board device (mainly processors and digital devices) with a badly designed DC path undergoes strong current drain variations.

One of the most 'mysterious' and dangerous parasitic phenomena, which we analyze in detail next, is known as 'injection locking', and produces several of the undesirable effects described above.

6.1.1 Injection Locking

Injection locking means: coupling (injecting) into an oscillator a signal of power much smaller than the natural oscillating signal, but close in frequency to the 3 dB bandwidth of the loaded resonator, causes the oscillator to lock up to the injected signal.

If locking occurs due to some transient effect, then the oscillator is momentarily 'pulled' out of frequency thereby generating unwanted spurious power. The coupling may be capacitive or inductive, by conduction or by radiation. Here we treat in detail the phenomenon in the realm of bipolar oscillators only. An extensive general treatment can be found in Razavi (2004).

Injection locking is sometime used intentionally and may be of advantage in specific designs, but more often it is a prime threat that greatly complicates layout and shielding requirements.

The mechanism in a bipolar oscillator using a feedback π topology is shown in Figure 6.1. As explained in Chapter 5, the input and output of the oscillating transistor are effectively disconnected.

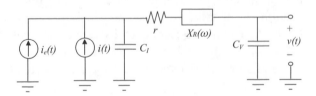

Figure 6.1   Injection locking mechanism

Here $i(t)$ represents the collector current, the voltage $v(t)$ is the RF base-emitter voltage, and $i_e(t)$ is the equivalent current induced by an external parasitic signal coupled (in whatever way) to the feedback network.

The feedback network is assumed to be resonant at frequency ω_0 with loaded quality factor $Q_L \gg 1$, thus only signals in the vicinity of ω_0 are considered.

We show now that an externally injected signal at nearby frequency $\omega \neq \omega_0$ will make the circuit oscillate at frequency ω, provided that certain conditions are met.

Assume that large oscillations at frequency $\omega \neq \omega_0$ indeed occur. With the fundamental base-emitter voltage given by

$$v(t) = V \cos \omega t = Re[V e^{j\omega t}] \qquad (6.1)$$

using the nonlinear bipolar transistor theory of section 5.4.1, and denoting by I_{dc} the transistor DC bias current while oscillating, we may write the fundamental collector current in the form

$$i(t) \approx -2I_{dc} \cos(\omega t) = Re[-2I_{dc} e^{j\omega t}] \qquad (6.2)$$

The current generated by the coupling of a parasitic signal of frequency ω and of power much smaller than the oscillator signal is

$$i_e(t) = -I_e \cos(\omega t + \phi) = Re[(-I_e e^{j\phi}) e^{j\omega t}], \ I_e \ll I_{dc} \qquad (6.3)$$

Recalling from section 5.2 that the transfer function of the feedback network is

$$Z(\omega) \approx -\frac{1}{\omega_0^2 C_I C_V r} \cdot \frac{1}{1 + j2Q_L \dfrac{\omega - \omega_0}{\omega_0}} \qquad (6.4)$$

then the oscillations must satisfy the complex-valued equation

$$V = (I_e e^{j\phi} + 2I_{dc}) \frac{1}{\omega_0^2 C_I C_V r} \cdot \frac{1}{1 + j2Q_L \dfrac{\omega - \omega_0}{\omega_0}} \qquad (6.5)$$

For $I_e \ll I_{dc}$ we may approximate

$$I_e e^{j\phi} + 2I_{dc} = I_e \cos \phi + 2I_{dc} + jI_e \sin \phi \approx 2I_{dc} + jI_e \sin \phi \qquad (6.6)$$

and setting

$$\Delta\omega \equiv \frac{\omega - \omega_0}{\omega_0} \qquad (6.7)$$

we may rewrite Equation (6.5) in the form

$$V \approx \frac{2I_{dc}}{\omega_0^2 C_I C_V r} \frac{1 + j(I_e/2I_{dc}) \sin \phi}{1 + j2Q_L \dfrac{\omega - \omega_0}{\omega_0}} \qquad (6.8)$$

Since the left side of Equation (6.8) is real, so too must be the right side.

Using Equation (6.7), we must satisfy

$$\frac{I_e}{2I_{dc}} \sin \phi = 2Q_L \frac{\Delta\omega}{\omega_0} \tag{6.9}$$

and since $|\sin \phi| \leq 1$, the 'locking range' where the oscillator locks up on the injected frequency is

$$\left|\frac{\Delta\omega}{\omega_0}\right| = \frac{I_e}{I_{dc}} \frac{1}{4Q_L} |\sin \phi| \leq \frac{I_e}{I_{dc}} \frac{1}{4Q_L} \tag{6.10}$$

and the angle difference between the oscillator output signal and the injected signal is

$$\phi = \sin^{-1}\left[\frac{I_{dc}}{I_e} 4Q_L \frac{\Delta\omega}{\omega_0}\right] \tag{6.11}$$

Substituting (6.9) into (6.8) we get the oscillation amplitude value, which is identical to the one previously obtained in section 5.4.2 for a non-injected oscillator, namely

$$V \approx \frac{2I_{dc}}{\omega_0^2 C_I C_V r} \tag{6.12}$$

We see that the locking range, defined by (6.10) and (6.11) depends both on the amplitude of the injected signal and on its offset from the network resonating frequency.

To get a feeling, assume a typical case: $f_0 = \omega_0/2\pi = 800$ MHz, $Q_L = 20$, and $I_e/I_{dc} = 1/10$. Substituting the above figures in (6.10) yields a lock range of ± 1 MHz.

6.1.2 Injection Pulling

Injection pulling occurs in transmitters working in intermittent mode, regardless of their modulation scheme. During power build up, drive-level dependent input and output impedances within the PA devices undergo momentary changes until they stabilize to their steady-state values. As a result, the phase of the RF signal building up in the final PA stage undergoes sudden variations. If the power amplifier radiates onto the oscillator that drives it, momentary VCO injection locking may occur resulting in frequency 'pulling' during power build up.

When the interference ceases, the oscillator has some offset with respect to the reference, and therefore the PLL is not at final state and must relock. As a consequence, the synthesizer starts a phase lock-up process as described in the transient analysis of Chapter 4. During the relock time, the transceiver may either transmit out of band or lose receiver sensitivity.

6.1.3 Remodulation

Remodulation is similar to injection pulling, but it occurs only if the transmitter carrier is amplitude-modulated, such as in QAM schemes.

As described before, PA radiation may produce injection locking with consequent frequency pulling (ground currents may also cause remodulation, as explained later). However, since the rate of change of the signal envelope is usually much faster than the synthesizer loop frequency, the PLL is unable to react although, *on average*, the VCO remains locked, just as described in Chapter 4 for fractional-N synthesizers.

However, the instantaneous VCO frequency is pulled back and forth, along with the changes in the transmitted power, which produces an additional parasitic FM modulation distorting the transmitted signal. Low-level remodulation may result in degradation of transmitter EVM or ACPR (Chapter 3).

6.1.4 Reverse Junction Breakdown

We saw in Chapter 5 that RF oscillations tend to reverse-bias bipolar junctions. In particular, in the case of silicon transistors, we saw that instantaneous junction breakdown will occur if the peak oscillating RF voltage v_{peak} satisfies

$$v_{peak} \geq \frac{V_{break} + 0.7}{2}[Volt] \qquad (6.13)$$

where V_{break} is the reverse junction breakdown voltage (usually of the order of 1.8 V). From that we conclude that whenever the base-emitter oscillation level reaches peak values above 1 V we should become concerned.

As pointed out when dealing with oscillators, when instantaneous reverse breakdown occurs, the value of the equivalent linear base-emitter loading resistance drops drastically (Clarke and Hess, 1978), thus loading the resonator and lowering the value of its loaded Q, which results in phase noise degradation. Thus, although under breakdown conditions the oscillator exhibits strong and 'healthy' oscillations, the SSB noise performance becomes inexplicably poor. We conclude that, upon completing an oscillator design the amplitude of the oscillations must be checked. Fortunately this is a very simple task, because, as we saw in Chapter 5, the amount of reverse DC biasing of the base-emitter junction as compared with the quiescent state (with no oscillations) is nearly equal to the peak oscillation voltage v_{peak}. This fact suggests a simple and accurate method for measuring the amplitude of oscillations as shown in Figure 6.2.

- Disconnect the loaded resonator from the base. The oscillator will stop working. Under quiescent condition measure the base-emitter quiescent DC voltage $V_{be}(quiescent)$.
- Reconnect the loaded resonator to the base. The oscillator will start working. Using a DC voltage meter, and through two 100 kΩ isolating resistors, measure the oscillating base-emitter DC voltage $V_{be}(oscillating)$.
- Compute the peak amplitude of RF oscillations

$$v_{peak} \approx V_{be}(quiescent) - V_{be}(oscillating) \qquad (6.14)$$

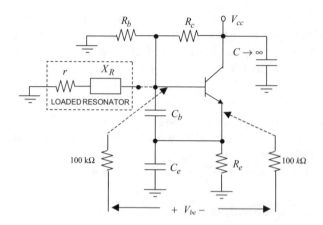

Figure 6.2   Measuring peak oscillation amplitude

6.1.5 Microphonics

Among parasitic effects, microphonics is one of the most serious and difficult to control. The phenomenon was first noted in FM receivers operating in voice mode: by 'tapping' on various locations on the transceiver body, the noise could be heard in the loudspeaker, just as if it was coming from a microphone.

In digital transceivers, microphonics does not show up in the loudspeaker, but produces severe signal-to-noise degradation in both receive and transmit mode, and may drastically impair bit error rate (BER) during vehicular operation.

The two main sources of microphonics are:

- Mechanical vibration reaching frequency-sensitive elements, such as the components used to construct the VCO resonator, including inductors, multi-layer chip capacitors, printed microstrips etc. Mechanical bending of these components changes their values thus producing frequency modulation of the VCO.
- Vibration of metallic objects in close proximity to resonator components or printed lines connected to the resonator, produces changes in the mutual capacitive/inductive coupling. Coupled objects are part of the VCO resonating system, so that a change in coupling produces a change in the VCO frequency. Metallic shields on the VCO and its neighborhood are a typical example.

Beside extreme care in layout and mechanical design, the only other cure for microphonics consists in designing a PLL with loop frequency greater than the highest vibration frequency. Such a PLL will react fast enough, thereby 'tracking out' the parasitic modulation, just as explained in section 1.3.3.1.

6.1.6 Ground Currents

When transmitting variable-amplitude signals, which is the case in most modern digital systems, the large PA current follows the amplitude variations. While the whole current flows to the PA through the positive DC line, the ground return path is spread over the printed circuit board (PCB).

Now assume that there is some physical distance between the ground of the VCO and the ground of the phase detector/loop filter of the synthesizer (Chapter 4). This physical gap is usually occupied by a copper ground plane, and is a part of the DC return path of the PA, and therefore part of the large variable current driven by the PA flows through it. Due to the resistance of the copper, the current variations generate corresponding voltage variations (at the modulation rate) that appear in series to the steering line of the VCO (Chapter 5).

Since the VCO has a sensitivity of many MHz/V, these parasitic voltages, although very small (of the order of magnitude of a fraction of a μV), produce significant FM modulation that will show up in the VCO spectral picture, and may exceed the required spectral mask (Chapter 3). This effect results in non-repeatable spectral purity performance of the transmitter even across the same production lot. Tightening various screws of internal shields or external boxes, adding or removing gaskets, or doing any other mechanical change, will not help. All these actions will modify the current distribution in the ground return path, resulting in unpredictable and non-repeatable spectral results.

In summary, once this phenomenon is in place, it has no cure apart from a major redesign of the layout, therefore, great attention should be used when setting the relative positioning of phase detector and VCO. To illustrate how serious this phenomenon is, let us get a feeling of its order of magnitude with a simple example.

Example: Assume that, due to QAM modulation, the PA exhibits current variations of 1 A peak, which, for the sake of simplicity, we approximate as a sinusoidal current at a symbol rate f_s, say

$$f_s = 14 \ kHz, \quad \omega_s = 2\pi f_s, \quad i(t) = cos(\omega_s t) \ [A] \tag{6.15}$$

Assume that the PCB is about one inch wide, and let us assume (again for simplicity) that the return ground path is such that the PA current distributes uniformly over the PCB width. Thus the current density per unit width i_D is

$$i_D(t) = i(t) \ [A/inch] \tag{6.16}$$

Assume that the gap between the ground of the VCO and the ground of the phase detector/loop filter is a copper strip of length d of 100 mil and width w of 15 mil. The thickness h of the copper in the PCB is 1.4 mil.

$$d = 0.1, \quad w = 0.015, \quad h = 0.0014 \ [inch] \tag{6.17}$$

With μ and σ respectively denoting the permeability and the conductivity of the copper, the skin depth δ (Collin, 1966) (depth of penetration) at 14 kHz is

$$\delta = \sqrt{\frac{2}{\omega\mu\sigma}} \approx 22 \text{ mil} \tag{6.18}$$

Thus, the current fully penetrates the copper layer (which is a best case, since the gap resistance turns out smallest).

The resistance of the gap R is about

$$R = \frac{d}{\sigma wh} \approx 3.2 \times 10^{-3} \ \Omega \qquad (6.19)$$

The peak current i_w crossing the width w is

$$i_w = max \ |w \cdot i_D| = 0.015 \ [A] \qquad (6.20)$$

Assuming that the VCO sensitivity K_V is

$$K_V = 10\frac{MHz}{V} \qquad (6.21)$$

the peak frequency deviation Δf is

$$\Delta f = K_V i_w R \approx 480 \ Hz \qquad (6.22)$$

and the FM modulation index β is

$$\beta = \frac{\Delta f}{f_s} \approx 3.4 \times 10^{-2} \qquad (6.23)$$

which, as seen in Chapter 5, results in a relative sideband level of

$$20 \ log \ (\beta/2) \approx -35 \ dBc \qquad (6.24)$$

The result of Equation (6.24) is unbearable under any circumstances.

6.1.7 Parasitic Poles and PLL Stability

We stated in Chapter 4 that a 'pure' second-order, type-two PLL cannot provide the required spur performance and a pre-integration capacitor must be added in the loop filter in order to clean up the high frequency portion of the phase detector error signal.

We also pointed out that the pre-integration capacitor creates a parasitic pole in the PLL transfer function. We showed that this parasitic pole can usually be neglected for the purpose of lock time computation. However, as a result of component variations, the additional pole, if not accurately controlled, may approach the right side of the plane in the Laplace domain. If this occurs, as we shall see shortly, stability may become marginal and if the pole crosses to the right side of the plane, stability is lost and PLL oscillations occur.

The general analytic computation of this phenomenon is not straightforward, since it requires solving a third-degree algebraic equation to find the PLL poles. The poles may be easily found for specific cases but a general simple analytic solution is not easy using traditional methods. This is why most engineers revert to computer simulation. However, the computer simulation masks the understanding of the phenomenon, as it gives only point-wise information without insight into the behavior. In the following we provide a general analytic solution of this problem by using a useful mathematical tool known as 'singular perturbation theory'.

Perturbation methods are very powerful when the solution of a problem depends on a small parameter ε, which we call 'a perturbation', *if* we know the solution of the 'unperturbed' (order-zero) problem, namely for $\varepsilon = 0$. In this case we can find an approximate solution of the perturbed problem at any desired accuracy $O(\varepsilon^n)$, $n = 1, 2 \ldots$ Here we bring a very concise and specialized overview of this theory applied to our problem. A deeper insight and tutorial can be found in Bender and Orszag (1978).

6.1.7.1 Perturbative Analysis

This example is a simple one, but it illustrates the power of perturbation methods. We repeat here the outcome of the equations for the transfer function of the loop filter of Chapter 4, and for the sake of simplicity we approximate $1 + \varepsilon \approx 1$, $\varepsilon \to 0$

$$H(s) = K_L \frac{s + \omega_L}{s} \cdot \frac{\omega_p}{s + \omega_p}, \, \omega_p = \omega_L/\varepsilon \longrightarrow \infty, \, \varepsilon \longrightarrow 0 \tag{6.25}$$

We also repeat the equation which describes, in the Laplace domain, the frequency offset behavior of the PLL following a change ΔN in the synthesizer divider value at $t = 0$

$$\Delta\Omega(s) = \frac{1}{s} \frac{K_\phi K_V H(s)}{Ns + K_\phi K_V H(s)} \Delta N \omega_{ref} \tag{6.26}$$

We take into account the effect of the pre-integration capacitor, by substituting (6.25) into (6.26), which yields

$$\Delta\Omega(s) = \frac{\alpha\omega_L(s + \omega_L)}{s[\varepsilon s^3 + \omega_L(s^2 + \alpha s + \alpha\omega_L)]} \Delta N \omega_{ref}$$

$$\alpha = \frac{K_\phi K_V K_L}{N} \in \mathbb{R}, \, \varepsilon \longrightarrow 0 \tag{6.27}$$

We now find a simple analytic expression for the poles of $\Delta\Omega(s)$ using singular perturbation theory, then, once the poles are known, we may find the PLL behavior just as we did in section 4.3.

For ease of understanding, we highlight the basic mathematical theory while proceeding with the actual solution of our problem, which is finding the roots of the denominator of Equation (6.27).

$$\varepsilon s^3 + \omega_L s^2 + \alpha\omega_L s + \alpha\omega_L^2 = 0, \, \varepsilon \longrightarrow 0 \tag{6.28}$$

First, we note that if $\varepsilon \equiv 0$, Equation (6.27) becomes the standard second-order PLL equation previously analyzed in section 4.3. In this case, we see that $\Delta\Omega(s)$ has only three poles. In contrast, for $\varepsilon \to 0$, but as long as $\varepsilon \neq 0$, $\Delta\Omega(s)$ has four poles. Therefore, for $\varepsilon \equiv 0$ the architecture of the solution of (6.28) is fundamentally different from the case where ε is very small but still non-vanishing. The explanation of this behavior is that the missing root tends to infinity as $\varepsilon \to 0$, thus, for this root, it is no longer valid to neglect $\varepsilon|s|^3$ as compared to the other terms in the limit when $\varepsilon \to 0$. When this happens we say that the perturbation problem is 'singular'.

A singular behavior shows up whenever the highest power in the equation is multiplied by some power of the small parameter ε. Of course, the other two roots computed in the unperturbed problem are indeed such that $\varepsilon|s|^3$ may be neglected in their vicinity, and we get the order-zero equation

$$s^2 + \alpha s + \alpha\omega_L \approx 0 \tag{6.29}$$

For the under-damped case, the complex conjugate roots of (6.29), namely $s_0^{(1)}$ and $s_0^{(2)}$, have been found to be

$$s_0^{(1)} = \gamma, \, s_0^{(2)} = \overline{\gamma}, \quad \gamma = \omega_n(-\xi + j\sqrt{1 - \xi^2})$$

$$\xi < 1, \, \xi = \frac{1}{2}\sqrt{\frac{\alpha}{\omega_L}}, \quad \omega_n = \sqrt{\alpha\omega_L} \tag{6.30}$$

To track the 'missing' root, we must estimate its magnitude as $\varepsilon \to 0$.

We should find a range of s where Equation (6.28) has at least two 'dominant' elements while all the others are negligible. When such dominant elements cancel each other, we call it a 'dominant balance'.

The dominant balance constitutes an approximation to Equation (6.28) valid in that neighborhood of s and yielding the missing root.

To find out if there exists some neighborhood of s where two elements in (6.28) constitute the dominant balance, namely, all other elements are negligible compared to them in that range of s, we must compare the orders of magnitude for all the relevant combinations of the elements of (6.28).

Suppose that εs^3 and $\omega_L s^2$ are dominant in magnitude for some range of s as $\varepsilon \to 0$, then

$$\varepsilon s^3 = O(s^2) \Rightarrow s = O(\varepsilon^{-1})$$

It follows that εs^3 and $\omega_L s^2$ are both $O(\varepsilon^{-2})$, since

$$\varepsilon s^3 = \varepsilon \cdot O(\varepsilon^{-3}) = O(\varepsilon^{-2}), \quad s^2 = [O(\varepsilon^{-1})]^2 = O(\varepsilon^{-2})$$

Substituting $s = O(\varepsilon^{-1})$, for instance, in $\alpha\omega_L s$, yields

$$\alpha\omega_L s = O(\varepsilon^{-1}) < O(\varepsilon^{-2}), \quad \varepsilon \longrightarrow 0$$

and of course

$$\alpha\omega_L^2 = O(1) < O(\varepsilon^{-2}), \varepsilon \longrightarrow 0$$

Thus, if $s = O(\varepsilon^{-1})$, the elements εs^3 and $\omega_L s^2$ not only have the same order of magnitude as $\varepsilon \to 0$, but they are also of the largest order of magnitude and thus they constitute the dominant balance. Indeed it is easy to verify that, if we suppose that some other combination is dominant, we get a contradiction.

If we suppose that εs^3 and $\alpha\omega_L s$ are dominant as $\varepsilon \to 0$ then

$$\varepsilon s^3 = O(s) \Rightarrow s = O(\varepsilon^{-1/2})$$

then εs^3 and $\alpha\omega_L s$ are both $O(\varepsilon^{-1/2})$, since

$$\varepsilon s^3 = \varepsilon O(\varepsilon^{-3/2}) = O(\varepsilon^{-1/2}), s = O(\varepsilon^{-1/2})$$

Substituting $s = O(\varepsilon^{-1/2})$ into $\omega_L s^2$ we get an inconsistent result because

$$\omega_L s^2 = O(\varepsilon^{-1}) > O(\varepsilon^{-1/2})$$

Thus εs^3 and $\alpha\omega_L s$ are not the largest elements.

Similarly, if we suppose that εs^3 and $\alpha\omega_L^2$ are dominant as $\varepsilon \to 0$ we get

$$\varepsilon s^3 = O(1) \Rightarrow s = O(\varepsilon^{-1/3})$$

Thus εs^3 and $\alpha\omega_L^2$ are both $O(1)$.

Substituting $s = O(\varepsilon^{-1/3})$ into $\omega_L s^2$ we get again an inconsistent result because

$$\omega_L s^2 = O(\varepsilon^{-2/3}) > O(\varepsilon^{-1/3})$$

so that εs^3 and $\alpha\omega_L^2$ are not the largest elements.

The next step is to introduce a change of variable into (6.28) in the form

$$s = \varepsilon^{-1} y \tag{6.31}$$

which yields the modified equation

$$y^3 + \omega_L y^2 + \alpha\omega_L \varepsilon y + \alpha\omega_L^2 \varepsilon^2 = 0, \ \varepsilon \longrightarrow 0 \tag{6.32}$$

Equation (6.32) represents a regular perturbation problem, since the highest power of y is not multiplied by ε.

Now we would like to find a solution that holds for *any* small value of ε. The standard technique consists of substituting a solution of the form

$$y = \sum_{n=0}^{\infty} y_n \varepsilon^n \tag{6.33}$$

and then requiring that the equation holds separately for each order of magnitude $O(\varepsilon^n)$.

For small values of ε, an error of $O(\varepsilon^2)$ is usually adequate, thus we substitute in (6.32)

$$y = y_0 + y_1\varepsilon \tag{6.34}$$

getting

$$(y_0 + y_1\varepsilon)^3 + \omega_L(y_0 + y_1\varepsilon)^2 + \alpha\omega_L\varepsilon(y_0 + y_1\varepsilon) + \alpha\omega_L^2\varepsilon^2$$
$$= y_0^3 + \omega_L y_0^2 + (3y_0^2 y_1 + 2\ \omega_L y_0 y_1 + \alpha\omega_L y_0)\varepsilon + O(\varepsilon^2) = 0 \tag{6.35}$$

Assuming $y_0 \neq 0$, we get two equations, one for each order of magnitude

$$O(1) : y_0^3 + \omega_L y_0^2 = 0 \Rightarrow y_0 = -\omega_L \tag{6.36}$$

$$O(\varepsilon) : 3y_0^2 y_1 + 2\ \omega_L y_0 y_1 + \alpha\omega_L y_0 = 0 \Rightarrow y_1 = -\frac{\alpha\omega_L}{3y_0 + 2\ \omega_L} = \alpha \tag{6.37}$$

where the right-hand side of the last equation was obtained by substituting (6.36) into (6.37). Thus the solution for y is

$$y = -\omega_L + \alpha\varepsilon + O(\varepsilon^2), \ \varepsilon \longrightarrow 0 \tag{6.38}$$

Substituting back (6.31), we finally get the additional root

$$s_p = -\frac{\omega_L}{\varepsilon} + \alpha + O(\varepsilon), \ \varepsilon \longrightarrow 0 \tag{6.39}$$

Note that (6.39) is valid for any small ε we may decide to pick, and it constitutes a generalized analytic solution for the additional root. Thus, $\Delta\Omega(s)$ has the form

$$\Delta\Omega(s) = \left(\frac{A}{s} + \frac{B}{s - \gamma} + \frac{C}{s - \overline{\gamma}} + \frac{B}{s - s_p}\right)\Delta N\omega_{ref} \tag{6.40}$$

Note that, as $\varepsilon \to 0$, s_p runs to infinity, but if ε becomes too large, then s_p may approach zero or even cross over to the right side of the complex plane, leading to instability.

Indeed recall that with L^{-1} indicating the inverse Laplace transform

$$L^{-1}\left[\frac{1}{s - \lambda}\right] = e^{\lambda t}u(t), \ \lambda \in \mathbb{C} \tag{6.41}$$

and, if $R_e[\lambda] > 0$, (6.41) becomes exponentially growing.

In order to guarantee that (6.40) yields a stable behavior, ε must be small enough. In particular, since in our simple case s_p is real, (6.30) and (6.39) imply that

$$s_p < 0 \Rightarrow \frac{1}{\varepsilon} > \frac{\alpha}{\omega_L} \Rightarrow \varepsilon < \frac{1}{4\xi^2} \tag{6.42}$$

In actual synthesizer designs, for various practical reasons, the topology of the loop filter together with the additional pre-integration/filtering actions often give rise to four poles in the expression for $\Delta\Omega(s)$, increasing the risk of instability.

In practice, since ξ depends on the (variable) division number N as well as on K_ϕ, K_V and K_L, which are subject to substantial variations over aging, temperature and components, it is customary to take

$$\varepsilon \ll \frac{1}{4\xi^2} \tag{6.43}$$

For instance, if we start with $\xi = 0.5$, then to be on the safe side it would be wise to keep $\varepsilon < 1/10$. In other words, referring to section 4.3.2, since $C_0 = \varepsilon C$, we should choose the pre-integration capacitor C_0 to be at least 10 times smaller than the loop filter capacitor C.

In the case of the fractional-N synthesizer with *MASH* topology described in section 4.2.4.4, we see that the capability to redistribute the reference noise towards the high frequency range is of great help as we can keep ε small and still obtain a good attenuation of the reference spurs.

6.2 Intercept Point and Spurious Responses

The 'intercept point' concept is a useful tool that provides a general characterization of nonlinear behavior of narrowband systems, and applies to both transmitters and receivers. In fact, the outcome of this section has been widely used in Chapter 2 to compute receiver specifications.

To understand the intercept point concept, consider a nonlinear system with an input signal consisting of several narrowband subcarriers

$$S_{in}(t) = \sum_{m=1}^{M} v_m \cos(\omega_m t + \theta_m), \quad v_m > 0 \tag{6.44}$$

where v_m, θ_m are either constants or narrowband signals (that can be considered constant in our context).

Note that $S_{in}(t)$ may be either the signal at the input of a receiver, or the signal to be transmitted at the input of an RF amplifier. In both cases, by representing $\cos(\omega_m t + \theta_m)$ in Euler form, the output of a nonlinear system may be written as

$$S_{out}(t) = \sum_{n=0}^{\infty} a_n S_{in}^n(t)$$

$$= \sum_{n=0}^{\infty} a_n 2^{-n} \left[\sum_{m=1}^{M} v_m (e^{j(\omega_m t + \theta_m)} + e^{-j(\omega_m t + \theta_m)}) \right]^n \tag{6.45}$$

where $\{a_n\}$ represent the nonlinearity of a receiver chain (LNA, mixers, IF amplifiers etc.) or a transmitter chain (drivers, power amplifier etc.).

The nth power in (6.45) yields real values $\{s_{n,i}\}$ of the form

$$s_{n,i} = 2Re[\gamma_{n,i}e^{j\omega_{n,i}t}], \gamma_{n,i} \in \mathbb{C}, i = 1, 2, \ldots$$

$$\omega_{n,i} = \sum_{m=1}^{M} k_{n,i,m}\omega_m, k_{n,i,m} \in 0, \pm1, \pm2, \ldots, \pm n \qquad (6.46)$$

where for each value of n, (6.45) implies that (6.46) must also satisfy

$$\sum_{m=1}^{M} |k_{n,i,m}| = n, |\gamma_{n,i}| = |a_n 2^{-n}| \prod_{m=1}^{M} v_m^{|k_{n,i,m}|} \qquad (6.47)$$

In the analysis that follows we assume that the amplitude of the coefficients $\{a_n\}$ drops fast enough so that if subcarriers of the same frequency ω have been generated for two different powers n_1, n_2 with $n_2 > n_1$, the overall product belonging to the higher distortion order is negligible. In other words, we assume that $n = n_1$ is the smallest distortion order capable of generating the frequency ω at an appreciable level. This assumption, although rather heuristic, is proven to be accurate in most actual situations (see the analysis of nonlinear amplifier behavior in Chapter 3).

Summarizing:

- $S_{out}(t)$ consists of a combination of subcarriers, each of amplitude proportional to *some* product of the individual amplitudes in $S_{in}(t)$.
- If there is a smallest possible order of distortion n for which one of the distortion products has a frequency ω that falls inside the band of interest and is dominant in power at that frequency, it will constitute an interfering signal of absolute amplitude V such that, for some positive constant α

$$V = \alpha \prod_{l=1}^{M} v_l^{|k_l|}, \quad \omega = \sum_{l=1}^{M} k_l\omega_l, \sum_{l=1}^{M} |k_l| = n, \alpha > 0 \qquad (6.48)$$

- If any one of the input subcarriers in (6.44) vanishes, then the spurious products itself vanishes as well. In other words, in order to generate a specific interference, *all* the input subcarriers corresponding to non-vanishing values of $\{k_l\}$ must have non-zero amplitudes $\{v_l\}$. Therefore, if only $m \le M$ values of k_l participate in the generation of some spur, we may write

$$V = \alpha \prod_{l=1}^{m} v_l^{|k_l|}, \quad \omega = \sum_{l=1}^{m} k_l\omega_l, \sum_{l=1}^{m} |k_l| = n, m \le M, \alpha > 0 \qquad (6.49)$$

- Some of the input subcarriers may have a fixed level (such as the local oscillator level in a mixer circuit), so they will produce only a frequency shift of the interference, but will not contribute to variations in its amplitude, while others will have varying amplitudes and will determine the strength of the received or transmitted interference.

Assume now without loss of generality that

$$k_1, k_2, \ldots, k_m \neq 0, \ k_{m+1}, k_{m+2}, \ldots, k_M = 0 \qquad (6.50)$$

and that only $v_1, v_2, \ldots, v_q, q \leq n$ may vary in amplitude, then if we increase each of them by the same factor, say $\lambda > 0$, in view of (6.49), $|k_1| + |k_2| + \cdots + |k_q| < n$, and the increased spurious signal amplitude $\tilde{V}$ will satisfy

$$\tilde{V} = V \prod_{l=1}^{q} \lambda^{|k_l|} = V\lambda^N, \ N = \sum_{l=1}^{q} |k_l| \leq n \qquad (6.51)$$

In dB notation, since $\{k_l\}$ are integers, the spur will increase (decrease) by

$$(\tilde{V}/V)|_{dB} = N\delta V, \delta V \equiv 20 \log(\lambda) = \lambda|_{dB} \qquad (6.52)$$

We note that, if each one of $v_1, v_2, \ldots, v_q$ increases by 1 dB, the interference itself increases by N dB. Thus, we say that the spurious product whose amplitude in given in (6.52) is of 'order N'. We stress that there is no direct correspondence between the order N of the spurious product and the order n of the distortion coefficient in (6.45), or the value of M in (6.44).

The order n of the distortion is determined by the number of subcarriers that participate in setting the spur frequency; the order N of the spur is determined by the number of signals that are allowed to vary in amplitude, while not all the M subcarriers present at the input must participate in generating the frequency of a specific spurious product. This point will be made clear in the forthcoming examples.

It is customary to set the variable amplitudes in (6.49) equal in level

$$v_1 = v_2 = \cdots v_q \equiv v \qquad (6.53)$$

which with N defined in (6.51) yields

$$V = \beta v^N, \ \beta = \alpha \prod_{l=q+1}^{m} v_l^{|k_l|} \qquad (6.54)$$

Since $v_{q+1}, v_{q+2}, \ldots, v_m$ and α are constant, then β is some positive constant.

Denoting (in dBm) the input signal power by p, the output signal power by P, and the output spur power by P_N we get

$$P = p + G, p = 10 \log(v^2)$$

$$P_N = 10 \log(v^{2N}) + 10 \log(\beta^2) = Np + G_N \qquad (6.55)$$

where G is the 'linear' power gain in dB, and $G_N = 20 \log_{10}(\beta)$.

The last equation in (6.55) describes a straight line with slope N, showing that if we increase the power of the 'fundamental' input signals by 1 dB, the power of the of the Nth order output spurious product increases by N dB.

Now, set $v_1, v_2, \ldots, v_q$ each at a level corresponding to a power p and measure the output spur power generated at the frequency ω in (6.49). We define as the *input intercept point of order N*, denoted by *IPNi* the value of p for which the above spurious power is equal to the power generated at the output by the 'linear' gain, when applying a single subcarrier of power p at the frequency ω to the input, i.e.

$$N \cdot IPNi + G_N = IPNi + G \Rightarrow G_N = (1 - N)IPNi + G \qquad (6.56)$$

then substituting back G_N into (6.55)

$$P_N = Np + (1 - N)IPNi + G \qquad (6.57)$$

We also define as the *output intercept point of order N*, denoted by *IPNo* the output power corresponding to *IPNi*, namely

$$IPNo = IPNi + G \qquad (6.58)$$

Using P in (6.55) and P_N in (6.57), the 'signal-to-distortion ratio' R (in dB) has the value

$$P - P_N = (N - 1)(IPNi - p) \ [dB] \qquad (6.59)$$

It follows from (6.59) that if *IPNi* is known, we can find the interference as a function of the input power of q interferers of equal intensity. Thus, *IPNi* fully characterizes the circuit behavior for a specific type of spur in a specific circuit. Again, one should bear in mind that in the intercept point context, 'input power' means the (equal) power of all the signals involved in generating a spur.

A feeling of the character of (6.59) can be gained from Figure 6.3. The following examples will make clear the above (rather confusing) statements.

6.2.1 Receiver Intermodulation Rejection (IMRN)

Before we proceed with the examples, let us find a general formula for computing receiver intermodulation rejection for any arbitrary order N.

In the following all the values are in dB or dBm. As in Chapter 2, *Sens* denotes the receiver sensitivity in dBm, and *CCR* denotes the co-channel rejection in dB. Using the results of section 2.4.1, and with reference to section 6.2 above, $IMRN \equiv (p - Sens)$ is the Nth order receiver intermodulation rejection, defined as the ratio (in dB) between the power level p for which the spur power P_N has the value

$$P_N = G + Sens + CCR \qquad (6.60)$$

and the receiver sensitivity itself.

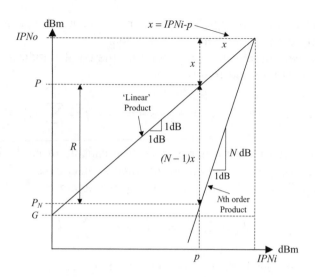

Figure 6.3 Nth order intercept point

Splitting $Sens = N\ Sens - (N-1)\ Sens$ in (6.60) and equating to (6.57) we get

$$IMRN = \frac{(N-1)}{N}(IPNi - Sens) + \frac{1}{N}CCR\ [dB] \qquad (6.61)$$

Example I: *IMR*3 of a mixer

The computation formula for *IMR*3 of a receiver has been given in section 2.4.1. Although several circuits such as LNA and IF amplifiers contribute to receiver intermodulation, the limiting factor for this specification usually lies in the non-linearity of the mixer(s).

We stress that, although the mixer intermodulation results in a third-order spur, the mechanism generating it lies in the fourth-order nonlinearity. We could have guessed this in advance, since one of the signals participating in the mixer operation is the local oscillator (LO), which has fixed amplitude.

Using the notation of section 2.4.1, the input signal consists of two carriers spaced Δf and $2\Delta f$ from the receiver center frequency f_R and of equal variable amplitude v, plus the local oscillator signal that has fixed amplitude v_{LO} and frequency $f_R - f_{IF}$

$$S_{in}(t) = v_1 \cos 2\pi f_1 t + v_2 \cos 2\pi f_2 t + v_{LO} \cos 2\pi f_{LO} t$$

$$f_1 = f_R + \Delta f, \ \ f_2 = f_R + 2\Delta f, \ \ f_{LO} = f_R - f_{IF} \qquad (6.62)$$

The offset Δf is large enough so that, unless there is a nonlinearity effect, both f_1 and f_2 fall outside the IF filter and are not detected. However we note that some nonlinearity produces the frequency combination

$$2f_1 - f_2 + f_{LO} = 2(f_R + \Delta f) - (f_R + 2\Delta f) - (f_R - f_{IF}) = f_{IF} \qquad (6.63)$$

The result is a spurious signal right at the IF center frequency, which will interfere directly with the desired receiver signal.

In order to check the mechanism generating the combination in (6.63), setting $f_3 \equiv f_{LO}$, we substitute into (6.49)

$$m = 3, \ k_1 = 2, \ k_2 = -1, \ k_3 = 1 \Rightarrow n = \sum_{l=1}^{m} |k_l| = 4 \qquad (6.64)$$

It follows that the smallest order of distortion capable of producing the spur is $n = 4$. However, only the first two out of three signals have variable amplitude, thus, setting $q = 2$ in (6.51), and summing up we obtain

$$q = 2, \ k_1 = 2, \ k_2 = -1 \Rightarrow N = \sum_{l=1}^{q} |k_l| = 3 \qquad (6.65)$$

So, this third-order spur is determined by a nonlinearity of order four.

Example II: *IMR3* **of a low-noise amplifier (LNA)**

The computation for the LNA is identical to the one for the mixer, except for the fact that there is no LO. The input signal is

$$S_{in}(t) = v_1 \cos 2\pi f_1 t + v_2 \cos 2\pi f_2 t$$

$$f_1 = f_R + \Delta f, \ f_2 = f_R + 2\Delta f \qquad (6.66)$$

and both subcarriers are allowed to vary in amplitude. In this case, (6.64) becomes

$$m = 2, \ k_1 = 2, \ k_2 = -1 \Rightarrow n = \sum_{l=1}^{2} |k_l| = 3 \qquad (6.67)$$

while (6.65) remains unchanged, namely

$$N = \sum_{l=1}^{2} |k_l| = 3 \qquad (6.68)$$

We see that for the LNA the same third-order spur is determined by the nonlinearity of third order as well.

6.2.2 Transmitter Intermodulation Distortion (IMDN)

Let us find a general formula for transmitter intermodulation distortion for any arbitrary order N as a function of *IPNi*.

Consider the input signal of section 3.2.5 consisting of two symmetrical sidebands with offset $\pm \Delta f$ with respect to the transmit frequency f_T

$$S_{in}(t) = v(\cos 2\pi f_1 t + \cos 2\pi f_2 t)$$

$$f_1 = f_T - \Delta f, \ f_2 = f_T + \Delta f, \ \Delta f \ll f_T \qquad (6.69)$$

whose average power, developed across a 1 Ω load has been shown to be

$$P_{Avg} \approx v^2 \tag{6.70}$$

According to the analysis of section 6.2, a dominant nonlinearity of order N in the transmitter chain will produce spurs at frequencies

$$f_N = m_1 f_1 + m_2 f_2 = (m_1 + m_2) f_T + (m_2 - m_1) \Delta f$$
$$|m_1| + |m_2| = N, \ |m_1|, |m_2| \geq 1 \tag{6.71}$$

and with amplitude

$$V = \beta v^N \tag{6.72}$$

A spur will fall in-band if, for some integer m

$$f_N = f_T + m \Delta f \tag{6.73}$$

Equating (6.73) to (6.71) and since $|m_1|, |m_2| \geq 1$ we get

$$\left. \begin{array}{r} m_2 + m_1 = 1 \\ (m_2 - m_1) = m \end{array} \right\} \Rightarrow m = 2m_2 - 1$$
$$\Rightarrow m = 2k + 1, \ k = 0, \ \pm 1, \pm 2, \ldots \tag{6.74}$$

Therefore (6.73) becomes

$$f_N = f_T \pm |2k + 1| \Delta f, k = 0, 1, 2, \ldots \tag{6.75}$$

Equation (6.75) shows that in-band spurs will be generated at both sides of the carrier, at frequencies with offset of odd multiples of Δf.

Since the spectral picture is symmetric with respect to f_T, it is enough to look at the sidebands of frequencies higher than f_T, for which k has non-negative values. From (6.74) and (6.51) it follows that for $k \geq 0$

$$\left. \begin{array}{l} |m_2| = \left| \dfrac{m + 1}{2} \right| = k + 1 \\[2mm] |m_1| = \left| \dfrac{1 - m}{2} \right| = k \end{array} \right\} \Rightarrow N = |m_1| + |m_2| = 2k + 1, \ k = 0, 1, 2, \ldots$$

$$\tag{6.76}$$

In view of (6.51), Equation (6.76) implies that a sideband spur with offset $(2k+1)\Delta f$ will be generated only by a nonlinearity of order $N \geq 2k+1$.

Assuming, as done before (see also Chapter 3), that the lowest order of distortion is dominant in amplitude, then by (6.72), the amplitude V_k of the kth sideband will be

$$V_k = \beta v^{2k+1} \tag{6.77}$$

Thus, the spur consisting of the sidebands at $\pm(2k+1)\Delta f$ is of order $2k+1$.

Denoting by p the input power P_{avg} in dBm

$$p = 10 \log (v^2/10^{-3}) \ [dBm] \tag{6.78}$$

and recalling the definition of *IMDN* given in section 3.2.5

$$IMDN = 10 \log \left(\frac{\text{power of fundamental sideband}}{\text{power of } [(N+1)/2]\text{-th sideband}} \right), \quad N = 2k+1 \tag{6.79}$$

A glance at Figure 6.3 reveals that the value of *IMDN* as a function of *IPNi* is given directly by (6.59), namely

$$IMDN = P - P_N = (N-1)(IPNi - p) \ [dB] \tag{6.80}$$

Example: Relationship between *IP3i* and the '1 dB compression point'
In section 3.2.5 we proved that

$$\frac{|c_k|}{|c_0|} v_{sat}^2 \approx \frac{\text{amplitude of } (k+1)\text{-th sideband}}{\text{amplitude of fundamental sideband}}, \quad k = 1, 2, 3, \ldots \tag{6.81}$$

In section 3.2.3 we showed that the 1 dB compression voltage satisfies

$$v_{sat} \approx 0.33 \left| \frac{c_0}{c_1} \right|^{\frac{1}{2}} \tag{6.82}$$

Substituting (6.82) for v_{sat} into (6.81) and setting $k = 1 (N = 3)$, we obtain

$$\frac{\text{amplitude of fundamental sideband}}{\text{amplitude of 2-nd sideband}} \approx 10 \tag{6.83}$$

Taking the square of both sides of (6.83) and computing the result in dB, it follows from (6.79) that

$$10 \log \left[\left(\frac{\text{amplitude of fundamental sideband}}{\text{amplitude of 2-nd sideband}} \right)^2 \right] = IMD3 \approx 20 \ dB \tag{6.84}$$

Substituting into (6.80) $N = 3$, $IMD3 = 20 \ dB$ and $p = p_{1 \ dB}$ yields

$$IP3i \approx p_{1 \ dB} + 10 \ [dBm] \tag{6.85}$$

which is the well-known rule of thumb stating that the intercept point of order three is about 10 dB higher than the 1 dB compression point. We observe that the *IP3i* is the imaginary point where the *IMD3* and fundamental products are intercepting. For a 'real-world' device, the fundamental and third-order products

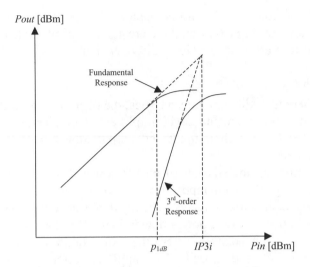

Figure 6.4   *IP3i* and the 'real world'

will eventually saturate before reaching this imaginary point as we can see in Figure 6.4.

6.2.3 Measurement of Input Intercept Points (IPNi)

IPNi cannot be measured directly but must be computed from *IMDN* and using Equation (6.80) to give

$$IPNi = p + \frac{IMDN}{N-1} \ [dBm] \qquad (6.86)$$

The setup is identical to the one used for measuring *IMDN* products in section 3.2.6. The measurement should be taken at the minimal input power level required to spot the relevant intermodulation product at the output, otherwise the results might be corrupted by the appearance of secondary PA saturation effects, feed-through phenomena or distortion in the measuring setup. To verify that the measurement is correct, we may decrease/increase the power of the input signals by 1 dB each and verify that the *IMDN* product at the output changes by *N* dB accordingly.

6.3 Parasitic Effects in Transmitters

Transmitters are prone to many parasitic effects. Some of them are due to the PA itself, others involve the whole transmitting chain, but virtually all of them show up in some form of instability, producing contamination of the output spectrum.

In many instances, transmitters that seem to work perfectly well upon production, start 'going crazy' after a while, whether under changes of environment, supply and RF load conditions, or for no apparent reason at all, and either oscillate, lose power or burn out. In fact, often there are early signs of the latent forthcoming troubles, and a little effort may help to discover them.

6.3.1 PA Instability

Due to their inherent nonlinear characteristics, the operation of power amplifiers is fundamentally different from low-level amplifiers. To understand this we should first recall how we define their input and output impedances, and why such a definition is needed.

The input and output impedances of an RF power device are mutually dependent, as well as dependent on input power, output power, DC supply and efficiency, and they cannot be reduced to a fixed physical lumped equivalent. In fact, the output impedance of a power device is usually defined as 'the conjugate of the optimum load impedance', where the word 'optimum' means 'to achieve maximal power output under specific DC supply and input power conditions'. A similar definition is used for the input impedance as well.

In practice, the input and output impedances of a power device must be found experimentally by sequentially rematching the input and the output ports, in a ping-pong sequence, until (hopefully) the process converges to some maximal output power with some maximal efficiency. Upon reaching optimal working conditions, we disconnect the power device from the matching circuitry, and measure the impedances seen at the connection points using a network analyzer.

The measured values are the result of the physical source and load impedances being transformed through the matching networks. Then the complex conjugates of the measured values are taken as the input and output impedances of the power device itself.

The process described above holds for the primary (large) signal being amplified by the device, for which we may define a set of 'large signal parameters'. However, what happens if, together with the primary *large* signal, we inject into the device input a secondary *small* parasitic signal?

It can be shown (Clarke and Hess, 1978) that in such a scenario, the gain seen by the small signal (the small-signal gain) *in the presence* of the large signal, might be considerably larger than the gain seen by the large signal (the large-signal gain). Moreover, the small-signal gain is dependent on the amplitude and shape of the (simultaneously coexistent) large signal. The small-signal input and output impedances also depend on the large signal, and thus, the phase shift experienced by the secondary signal going through the power device, depends on the primary signal too. Since the primary signal may change in shape and power, the small-signal behavior cannot be effectively controlled. Thus, in parallel to the large-signal system, no matter how accurately designed, there exists a small-signal system with unpredictable large gain, and unpredictable phase shift.

If a parasitic feedback path exists between the PA output and any of the interstage inputs of the transmitter chain, this is the perfect recipe for generating erratic oscillations. Unfortunately, parasitic feedback paths do exist, often resulting from improper layout, poor grounding, insufficient shielding, or lack of decoupling on the DC supply lines.

The mechanisms producing potential instability are so many that we won't even try to discuss them here. What we do instead, is to provide the reader with effective means for determining *whether* there is a potential instability danger in the transmitter chain, by looking for phenomena that we denote as spectral 'bumps'.

6.3.2 Spectral 'Bumps'

The spectral picture of a transmitted carrier, when the transmitter is free from instability, should look as in Figure 6.5. The noise shape seen away from the carrier has the shape described by Leeson's equation (Chapter 5) and decays at the rate of 6 dB/octave until it reaches the noise floor. If everything is okay, no odd shapes, spikes or bumps should be seen.

If, due to the reasons mentioned in the previous section, low-level positive feedback patterns are generated, then, in certain frequency bands, the small-signal gain becomes large and we may be able to see these bands appearing as 'bumps' in the spectral picture as shown in Figure 6.6.

These bumps are 'white' noise, which has been highly amplified in the frequency bands where the positive feedback induces large values of gain. The

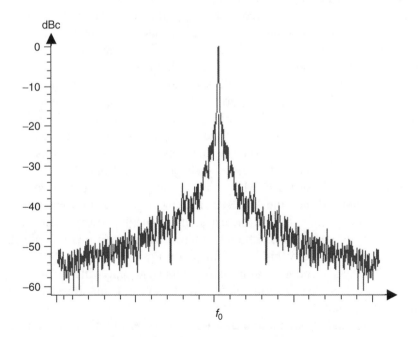

Figure 6.5 Instability-free spectrum

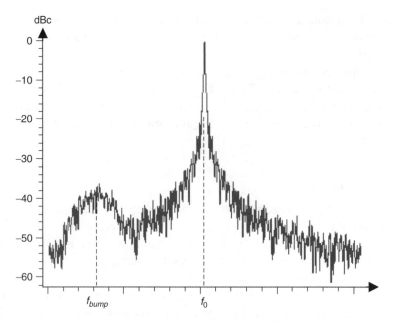

Figure 6.6   Strong spectral bump

bumps are also characterized by a relatively wide bandwidth, since the circuitry involved (usually the matching circuits) has low Q.

A transmitter exhibiting a bump of the type shown in Figure 6.6 has good chances of oscillating near the frequency f_{bump} following changes in DC supply, environmental conditions, aging, or antenna mismatch. Oscillations show up more often at low temperatures ($-60\,^\circ$C) and high DC voltages because under these conditions the small signal gain of transistors increases.

Not every bump is visible by simple spectral inspection, as its amplitude may be below the noise floor of the measuring system, and since, as mentioned before, the bumps are essentially amplified white noise, narrowing the bandwidth of the IF filter in the spectrum analyzer will not make them 'pop up', since their amplitude will be reduced by the same amount as the noise floor level. For this reason we need a more proactive method to discover latent spectral bumps.

The idea is straightforward: while transmitting at rated power, observe the output after 'notching out' the large carrier, and inject, at the beginning of the relevant transmit chain, a low-level test signal sweeping over whatever band is required. Since the test signal, although small, can be made much larger than the thermal noise, it will make 'hidden' bumps pop up.

Figure 6.7 shows a hidden bump at frequency f_{hidden}, which cannot be seen with the bare eye in the spectral picture of Figure 6.6, but reveals itself upon injecting a test signal.

Coupling a secondary signal to a transmitting power amplifier, without causing damage to the measuring equipment, and while observing relatively small signal

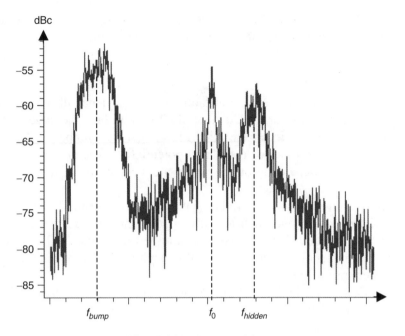

Figure 6.7 'Hidden' spectral bump pop-up

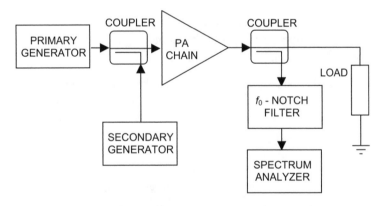

Figure 6.8 Spectral bumps – measurement setup

levels at the output, is not straightforward, and is accomplished with the setup of Figure 6.8.

6.4 Parasitic Effects in Receivers

Receivers suffer from self-generated disturbances, some due to nonlinear effects and others to receiver circuitry, as well as interferences due to associated transmitters, processors and digital circuits, local clocks, switching DC converters etc.

The variety of possible parasitic disturbances is unlimited. Here we explain and analyze in detail some of the most common and devastating.

6.4.1 Able–Baker Spurs

The 'Able–Baker' spurs are a general family of IF-frequency dependent quieters generated by mixer nonlinearity (Cijvat et al., 2002). In fact, the larger the distortion order in the nonlinearity of the mixer, the larger the number of Able–Baker spurs. However, although all of them may potentially cause desense, only a few eventually show up, depending on the actual architecture and construction of the receiver.

Since most of the many Able–Baker spurs are only potentially damaging, and we cannot know which ones will actually appear, they are analyzed in order to choose an IF frequency that will likely minimize their number. The target is to choose the IF frequency so that, ideally, all the spurs will fall outside the receiver RF front filter (preselectors).

Any combination of the LO and an RF signal that satisfies Equation (6.87), i.e. produces a mixer output at the IF frequency, is a potential Able–Baker spur.

$$\pm m f_R \pm n f_{LO} = \pm f_{IF} \tag{6.87}$$

where:

m denotes the mth harmonic of the received frequency f_R
n denotes the nth harmonic of the LO frequency f_{LO}
f_{IF} denotes the IF frequency of the receiver

Here we show the case of a receiver that has been designed to operate at low-side injection (the higher side injection analysis is similar). For low-side injection

$$f_R > f_{LO} > f_{IF} > 0 \tag{6.88}$$

In view of (6.88), Equation (6.87) becomes

$$f_R = \frac{n f_{LO} + f_{IF}}{m} \tag{6.89}$$

Every frequency f_R satisfying (6.89) is a potential Able–Baker spur. The image ($m = -1, n = -1, f_{LO} - f_{IF}$) and half-IF ($m = 2, n = 2, f_{LO} + f_{IF}/2$) are examples of Able–Baker spurs.

6.4.2 Self-Quieters

The name 'self-quieters' is a legacy from the world of analog FM receivers. It is a well-known outcome of FM theory (Schwartz, 1990) that, with no received signal at the antenna port, a strong 'white-noise' like signal appears at the output of the FM detector. In analog FM receivers operating in voice mode, this signal results in a strong noise coming from the loudspeaker.

If a parasitic unmodulated carrier signal reaches the receiving chain, whether from the antenna or directly into the IF lineup, due to the FM capture effect, the noise is greatly reduced, namely, the loudspeaker becomes 'quiet'.

One of the 'sanity tests' of FM receivers is checking that the white noise in absence of reception is indeed at the expected (strong) level.

If some interferer self-generated by the receiver causes a reduction of the detector output noise with no signal applied, this interfering signal is denoted by the name 'self-quieter'. Self-quieters may show up on one or more channels at once, and their presence may be difficult to detect. For instance, if only one channel is quieted, a receiver operating in frequency-hopping mode will suffer from little performance degradation, while a single-channel receiver will suffer from a substantial loss of sensitivity up to a complete paralysis.

The level of self-quieters is usually unstable, and may vary dramatically over time and environmental conditions. Their existence indicates some inherent marginality in the design, whether in shielding, DC filtering etc. Thus, no matter how small a self-quieter may be, RF designers must be worried about and take every possible measure to remove it.

The presence of a self-quieter can be confirmed by checking the level of reception. All modern receiver components have a function known as RSSI (received signal strength indicator), which gives a measure of the received signal. If the RSSI shows reception while no signal is applied, this is a confirmation of the existence of a self-quieter.

In looking for a self-quieter, it is important to understand whether it is being received from the antenna, or via some other mechanism such as conduction through DC lines, or direct radiation between on-board components and lines.

Therefore the RSSI should be monitored first with the antenna connected, and once again, with the antenna port terminated in a dummy load. In general, the case where the self-quieter is received by the antenna is the simpler to correct, since the mechanism is likely to be direct radiation from the radio body in the neighborhood of some high-level oscillator, and better shielding and grounding may suffice.

Self-quieters may result from a variety of sources:

- Wrong choice of receiver architecture. For instance picking up some clock frequency, in the receiver chain or in the digital circuits, whose harmonics fall close to one of the receiver channels, or, much worse, close to the IF frequency (at which the receiver-lineup gain may be very high).
- Data and address busses of microprocessors, memories and other fast digital subsystems produce both CW and wideband white-like noise. If their layout runs close to the receiver lineup, broadband interferers may result in sensitivity loss over many channels. In the CW case, the interference may be mitigated by

using a technique known as 'dithering', which consists in FM-modulating the crystal that generates the interfering clock using a fast pseudorandom sequence. The modulation causes the spectra of the clock harmonics to spread out (just as in CDMA), thus the interference power per unit bandwidth is reduced. If the receive bandwidth is fixed, the interfering power into the receiver will drop accordingly.

- The harmonics of DC switching regulators may fall in the receiving or IF band. Cases have been observed where the 400th harmonic of a DC switching regulator caused severe quieting.
- Another frequent cause of self-quieting due to improper design, lies in instability of the receiver LNA or of some stage in the IF amplifier chain. Such instability produces low-level oscillations in the vicinity of the receiver frequency, and may cause severe degradation in sensitivity performance.

6.4.3 Doppler Blocking

Doppler blocking is an interesting effect that has been observed to cause problems in DCR receivers (Chapter 2). In every receiver there is some leakage of the LO at frequency f_{LO} through the antenna port. The LO signal is radiated, hits nearby metallic objects (targets), is returned to the antenna and received just as an on-channel signal.

Since the LO is a non-modulated carrier, if the metallic target is at rest, the received signal will be right at LO frequency, will be decoded as DC, and will be rejected by the zero-IF filter. However, if the target has velocity v, the well-known Doppler effect (Skolnik, 1990) will make the received signal shift in frequency by a 'Doppler shift' f_d

$$|f_d| = 2 f_{LO} |v|/c \qquad (6.90)$$

where c is the speed of light. With this frequency shift, the signal will no longer be rejected by the DC notch, and will appear as an on-channel signal.

Since the reflection is from nearby objects, the strength of the reflected carrier may be stronger than the desired signal, which will cause momentary blocking of the receiver.

Effects of the above kind have been observed when shaking a bundle of keys in the vicinity of the antenna. As an example, at a walking speed of 3 m/sec, and with $f_{LO} \approx 2$ GHz, then $f_d \approx 40$ Hz.

6.4.4 Chopper Noise

'Chopper noise' is an annoying effect related to TDMA digital transceivers operating in digital voice mode. In spite of the fact that such transceivers are not analog, their acoustic accessories ultimately deliver analog signals.

The transmitted RF power radiates onto the acoustic accessories or onto the analog portion of the audio amplifiers and the associated DC supply, and is rectified in the bipolar junctions of the components. Since TDMA transmitters usually work in slow-rate pulses, the rectified voltages are heard in the local loudspeaker or captured by the microphone and heard by the far side as annoying 'clicks', similar to the noise of a chopper.

The only cure for chopper noise is a careful RF bypassing of low-frequency bipolar components, using small capacitors whose values are chosen so as to be in serial resonance at the transmitted frequency.

6.5 Specific Absorption Rate (SAR)

SAR is a relatively new safety specification, defining the amount of RF power, which is allowed to dissipate in various parts of the human body by electronic and communication hand-carried equipment. Usually SAR is specified in units of watt/kg, and the actual limit is very dependent on the specific part of the body, the frequency, the type of equipment and the country regulations.

The measurement of SAR requires very sophisticated equipment and systems simulating the human body, referred to as 'phantoms'. The actual measurement is carried out by checking the amount of heat dissipating in the phantom, which often consists of a bath of saline liquid.

Fortunately, accurate simulations of the RF dissipation are possible, using detailed body models, and finite-difference time-domain (FDTD) methods capable of solving Maxwell's equations within the complex electromagnetic system including the human body, as well as radio equipment with its associated detailed mechanical design (Kunz and Luebbers, 1993).

Other mathematical methods, such as the widely used 'Moment Method' (Wang, 1991), are not applicable in the human body case, because, due to the huge number of body elements that must be included, we end up with a prohibitive running time. Even with the FDTD approach, a full body simulation on a powerful multiprocessor parallel computer may last several hours.

Sometimes, a useful SAR estimate can be obtained using a (relatively) simple analytic computation, modeling the human body as an infinite two-layer fat and muscle biological cylinder with the following characteristics:

- Fat layer: $\varepsilon_r \approx 6$, $\sigma \approx 0.08$ [(ohm-m)$^{-1}$]
- Muscle layer: $\varepsilon_r \approx 42$, $\sigma \approx 1.8$ [(ohm-m)$^{-1}$]

where ε_r is the permittivity relative to vacuum.

A thorough insight into analytic electromagnetic modeling of human body may be found in Siwiak (1995).

Several good software packages are available for SAR computation, and they are extremely useful in the early design stages.

Many safety guidelines have been published, and accurate simulations and measurements of SAR are an absolute requirement before any equipment may be safely operated. As a general guideline: SAR is very strongly dependent on the physical proximity between the equipment and the human body, thus, when carrying out the conceptual design of a product, it is necessary struggle for the largest possible physical separation. In fact, keeping a physical distance of a few centimeters, at least, between the antenna and the human body, is essential in controlling SAR values.

Bibliography

Abramovitch, Milton and Stegun, Irene (1970). *Handbook of Mathematical Functions*. Dover Publications.

Balanis, Constantine A. (1997). *Antenna Theory*. John Wiley & Sons, Ltd.

Bender, Carl M. and Orszag, Steven A. (1978). *Advanced Mathematical Methods for Scientists and Engineers*. McGraw-Hill.

Chui, Charles K. (1995). *An Introduction to Wavelets*. Academic Press.

Cijvat, Ellie *et al.* (2002). Spurious mixing of off-channel signals in a wireless receiver and the choice of IF. *IEEE Transactions on Circuits and Systems – II, Analog and Digital Signal Processing*, **49**(8).

Clarke, K. K. and Hess, D. T. (1978). *Communication Circuits: Analysis and Design*. Addison-Wesley.

Collin, R. E. (1966). *Foundation for Microwave Engineering*. McGraw-Hill.

Crols, Jan and Steyaert, Michel S. J. (1995). A single-chip 900 MHz CMOS receiver front end with a high-performance low-IF topology. *IEEE Journal of Solid State Circuits*, **30**(12).

Crols, Jan and Steyaert, Michel S. J. (1998). Low-IF topologies for high-performance analog front ends of fully integrated receivers. *IEEE Transactions on Circuits and Systems – II: Analog and Digital Signal Processing*, **45**(3).

Dawson, Joel L. and Lee, Thomas H. (2004). Cartesian feedback for RF power amplifier linearization. *Proceedings of the 2004 American Control Conference*, Boston. June 30–July 2.

Dixon, Robert C. (1994). *Spread Spectrum Systems with Commercial Applications*. John Wiley & Sons, Ltd.

Fontana, Robert J. (2004). Recent system applications of short-pulse ultra-wideband (UWB) technology. *IEEE Transactions on Microwave Theory and Techniques*, **52**(9).

Gailus, P. and Charaska, J. (2002). 150/400/800/1900 MHz low noise Cartesian feedback IC with programmable loop bandwidth. *Proceedings of the 28th European Solid-State Circuits Conference*, September 24–26.

Gohberg, Israel and Goldberg, Seymur (1981). *Basic Operator Theory*. Birkhäuser.

Gross, Frank B. (2005). *Smart Antennas for Wireless Communications*. McGraw-Hill.

Heiter, George L. (1973). Characterization of nonlinearities in microwave devices and systems. *IEEE Transactions on Microwave Theory and Techniques*, **NTT-21**(12).

Jordan, Edward C. (1989). *Reference Data for Engineers: Radio, Electronics, Computer and Communications*. Howard W. Sams & Company.

Krauss, Herbert L., Bostain, Charles W. and Raab, Fredrick H. (1980). *Solid State Radio Engineering*. John Wiley & Sons, Inc.

Kunz, Karl R. and Luebbers, Raymond J. (1993). *The Finite Difference Time Domain Method for Electromagnetics*. CRC Press.

Lyman, R. J. and Wang, Q. (2002). A decoupled approach to compensation for nonlinearity and intersymbol interference. International Telemetering Conference, USA.

Millman, J., and Halkias, C., (1972). *Integrated Electronics*. McGraw-Hill.

Papoulis, Athanasios (1991). *Probability, Random Variables, and Stochastic Processes*. McGraw-Hill.

Perrott, Michael H. and Trott, Mitchell D. (2002). A modeling approach for $\sum -\Delta$ fractional-N frequency synthesizers allowing straightforward noise analysis. *IEEE Journal of Solid State Circuits*, **37**(8).

Proakis, John G. (1983). *Digital Communications*. McGraw-Hill.

Raab, Fredrick H. *et al*. (2003). RF and microwave power amplifier and transmitter technologies – part 2. *High Frequency Electronics*, July.

Raab, Fredrick H. *et al*. (2003). RF and microwave power amplifier and transmitter technologies – part 4. *High Frequency Electronics*. November.

Razavi, Bezhad (2004). A study of injection locking and pulling in oscillators. *IEEE Journal of Solid State Circuits*, **39**(9).

Rohde, Ulrich L. (1997). *Microwave and Wireless Synthesizers*. John Wiley & Sons, Inc.

Rozen, Kennet H. (2003). *Discrete Mathematics and its Applications*. McGraw-Hill.

Runkle, Paul *et al*. (2003) DS-CDMA: the modulation technology of choice for UWB communications. IEEE Conference on Ultra-Wideband Systems and Technologies.

Saberinia, Ebrahim and Tewfik, Ahmed H. (2003). Multi-user UWB-OFDM communications. IEEE Pacific Rim Conference on Communications, Computers and Signal Processing.

Saleh, Adel A. M. (1981). Frequency independent and frequency dependent nonlinear models of TWT Amplifiers. *IEEE Transactions on Communications*, **COM-29**(11), 1715–20.

Schwartz, Mischa (1990). *Information, Transmission, Modulation and Noise.* McGraw-Hill.

Siwiak, Kazimierz (1995). *Radiowave Propagation and Antennas for Personal Communications.* Artech House.

Sklar, Bernard (2001). *Digital Communications, Fundamental and Applications.* Prentice-Hall.

Skolnik, Merrill (1990). *Radar Handbook.* McGraw-Hill.

van Nee, R. and de Wild, A. (1998). Reducing the peak-to-average power ratio of OFDM. Vehicular Technology Conference.

Vidojkovic, Vojkan, van der Tang, Johan and van Roermund, Arthur (2001). Low-IF receiver planning for the DECT system. Department of Electrical Engineering, Eindhoven University of Technology.

Wang, Johnson J. H. (1991). *Generalized Moment Methods in Electromagnetics.* John Wiley & Sons, Ltd.

Yu, C. and Ibnkahla, M. (2005). Exact symbol error rate and total degradation performance of nonlinear M-QAM fading channels. *Proceedings of IEEE International Conference on Acoustics, Speech, and Signal Processing* (ICASSP '05), **3**, 18–23.

Index

Wireless Transceiver Design   Ariel Luzzatto and Gadi Shirazi
© 2007 John Wiley & Sons, Ltd